ROUTLEDGE SERIES ON SOVIET (RUSSIAN) MILITARY THEORY AND PRACTICE

IF WAR COMES TOMORROW?

IF WAR COMES TOMORROW?

The Contours of Future Armed Conflict

GENERAL
MAKHMUT AKHMETOVICH GAREEV

Translated by
Yakov Vladimirovich Fomenko

Edited and with an Introduction by
Jacob W. Kipp

Routledge
Taylor & Francis Group
LONDON AND NEW YORK

First Published in Russia by
Vladar Publishing House, Moscow, 1995
First English Edition Published in 1998
in Great Britain by
Routledge
2 Park Square, Milton Park, Abingdon, Oxon, OX14 4RN
711 Third Avenue, New York, NY 10017, USA

Routledge is an imprint of the Taylor & Francis Group, an informa business

British Library Cataloguing in Publication Data:

Gareev, Makhmut Akhmetovich
If war comes tomorrow? : the contours of future armed conflict. – (Cass series on Soviet (Russian) military theory and practice ; no. 7)
1. Military art and science 2. Military policy
I. Title
355'.03
ISBN 0-7146-4801-9 (cloth) 0-7146-4368-8 (paper)

Library of Congress Cataloging-in-Publication Data

Gareev, M. A. (Makhmut Akhmetovich)
[Esli zavtra voĭna? English]
If war comes tomorrow? : the contours of future armed conflict / Makhmut Akhmetovich Gareev ; translated by Yakov Vladimirovich Fomenko ; with an introduction by Jacob W. Kipp.
p. cm. — (Cass series on Soviet (Russian) military theory and practice)
Includes index.
ISBN 0-7146-4801-9. — ISBN 0-7146-4368 8 (pbk.)
1. World War III. 2. Military art and science—Forecasting. 3. International relations. 4. Military history—Philosophy. I. Title. II. Series.
U21.2.G36513 1998
355.4'8—dc21 97-32218
CIP

Typeset by Vitaset, Paddock Wood, Kent

Publisher's Note

The publisher has gone to great lengths to ensure the quality of this reprint but points out that some imperfections in the original may be apparent

CONTENTS

LIST OF ILLUSTRATIONS

PREFACE

There is no lack of accusation that the military are preparing for the last war. This catch-phrase, by the way, was coined by a well-known politician only after the war, when many things had become obvious. Everybody understands that it would be more prudent to orient military structuring and war readiness towards the future. But when it is hard to draw appropriate conclusions from the past, it is even harder to foresee the future. History knows of many accurate predictions regarding certain aspects of future wars, but nobody has yet succeeded in foreseeing the nature of any forthcoming armed struggles in their totality.

And still this labor of Sisyphus goes on. Prognostications concerning the causes of war and the nature of armed struggle have become more topical nowadays, and not only with regard to defense issues, but first and foremost in order to prevent wars and armed conflicts. It is better to err in prediction than to yield to the impossibility of foreseeing the progress of military affairs. And if only someone venturing to resolve the problems of the future would manage to infer at least a small detail, or a trend, the number of oversights would be smaller.

Profiling future conflict, one is tempted to chastise, as usual, everything obsolete and currently exploited; jumping into the world of mysterious and thrilling technologies, and unheard of ways of their utilization. But just as throughout history dialectical consistency was preserved in the development of military affairs regardless of turbulent revolutionary changes; so there are no grounds for expecting the times to be out of joint. Therefore, researchers should balance their daring forecasts, because normally nothing jumps up out of nowhere. In the past, miscalculations were common, due to one-sided approaches and extreme assessments, and some theoreticians maintained that war could be won by tanks and aircraft alone, or by following one method of fighting, forgetting that a diversity of ways and means is characteristic of any given reality. Even staking on such an 'absolute' weapon as the nuclear one failed expectations.

Envisaging newly developing methods of war has always been one of the most important tasks of military art. Laying the theoretical foundations is now even more complicated as the qualitative disparity increases between weapons of the last war and those of the future wars.

In the sphere of rapid scientific progress in military affairs one has to make decisions without the necessary amount of relevant experience. War games in times of peace can only be of limited use: it is impossible to simulate everything, particularly in the nuclear arena.

In history, after any great war (Franco–Prussian, Russo–Japanese, World War I), it was often stated that the current military art was outdated. But the next war, though introducing new methods, still preserved many old ways. There has never yet been a war that totally negated all former gains of military art.

Neither current experience nor superficial knowledge of the future are requisites, rather foresight depends on deep-rooted, often-concealed directions and tendencies, which possess the ability to evolve into completely new forms. No war experience is without significance. And along with this, the next war tends not to resemble the previous one, but throws up progressively more novelties. This factor contains within itself the seeds of new potential.

It is important that both sides of this contradictory dialectical process are considered by scientific research. As the Polish philosopher Leszek Kolakowski said, 'We need to remember two things: first, if not for new generations' rebellions against old traditions, we would have been still living in caves; and second, if there had only been those rebellions, we would have been in the caves again.'[1]

Thus military science should rebel against obsolete views, but at the same time avoid leaving in our modern caves anything that might help shed light on the future.

As the Wall Street journal once put it, 'the best Pentagon analysts worry not about Bosnia, Haiti or North Korea, but about timely apprehensions of possible cardinal changes of war fighting methods in the coming decades'.[2] This is also a burning issue for specialists in Russia and other countries.

I define armed struggle as part of war – war being a complex social and political phenomenon embracing all spheres of life in the nations engaged, with diverse ways of fighting the enemy including industrial, political and psychological means, with armed struggle being paramount among them.

That is why in order to answer the question of what armed struggle will be like in the future, one needs to know to what extent the political and social essentials might change in the new military-political environment, and what the main causes of war are likely to be.

The ways and means of fighting wars will still be greatly affected by social, political and industrial factors; new discoveries in weapons technology; and changes in societies' mentality with regard to war and the methods of manning them. An analysis of all these factors in the aggregate can more or less precisely outline tendencies and opportunities in the development of war-fighting methods. In doing this, and understanding the extensiveness of the subject, I intend to limit the scope of the analysis of future warfighting methods only to those which look to acquire definite new facets, and which are scantily surveyed in recent military publications.

The thoughts presented in this book are not only the result of my personal reflections and research, but emerged in the course of discussions, meetings and debates at scientific conferences, which took place at the Military Academy of the General Staff, the Center of Military and Political Research of the General Staff, the Academy of Military Sciences, the department of Geopolitics and Security of the Academy of Natural Sciences of the RF (Russian Federation), and at the Russian Institute of Strategic Research; and have been influenced by meetings and joint discussions at NATO headquarters and research centers, the London International Institute of Strategic Studies, the American Marshall Center (in Munich), German–Russian veterans' organizations, the US National Defense University, the American Space Transportation Association headed by Daniel O. Graham and Edward Teller, the American training centers at Ft. Leavenworth, Fort Sill and Benning, Knox, and the department of defense of Syria presided over by General M. Tlass. Such meetings and discussions almost never directly contribute to cognition of the truth, but as a rule they table new questions and fresh ideas fruitfully, and correlate opinions.

I express my gratitude to ONEKSIM-Bank, and the Russian Fund of Foreign Economics Policy and personally to Alexander Ivanovich Ageev, Sergey Anatoliyevich Batchikov, and Sviatoşlav Yurievich Rybas.

M.A.G. 1995

NOTES

1. Cited from *Ot strategii oborony k politike mira. Germanskie vooruzhennye sily v period peremen* (Moscow: Nauka Publications, 1993) p. 87.
2. *Kompas*, 1994, No. 119, pp. 3–7.

The military have been always accused of making preparations for past wars. And it is true, that in spite of some accurate predictions, almost nobody managed to foresee the character of future war. Nevertheless, this labour of Sisyphus goes on. For it is better to err in forecasts, than yield to the idea that it is impossible to foresee the future of military affairs.

INTRODUCTION

THE CONTOURS OF FUTURE ARMED CONFLICT AND THEIR INTERPRETER: Implications for National and International Security Policy

JACOB W. KIPP

THE BOOK

The appearance of General Makhmut Akhmetovich Gareev's book on the contours of future armed conflicts is timely. General of the Army M.A. Gareev, Russian Army (retired), has authored a well-crafted study of those contours, focusing on the political, economic, social, and technological trends which will affect the evolution of military art of the next ten to fifteen years. The book is a well-argued case of the need for foresight to guide military doctrine, force structure, and force development. Gareev admits that the task of foresight is like 'the labor of Sisyphus' but sees an ongoing process of forecasting as a necessary element of military science.

Gareev focuses upon the application of foresight to resolve defense problems with an eye toward preventing wars and armed conflicts. He explicitly warns against two dangers. The first is the assumption that forecasting the development of military art is so complex that no forecast will have any real value. The second is the danger of turning a forecast into an advocacy exercise for a particular weapons system or military service, in which that weapon's proponents transform it into an absolute solution to all defense problems, and use it to legitimize the allocation of resources in the bureaucratic struggle for funding. Gareev's categories are those of Soviet military science and his primary focus is upon those changes which will bring about a break in the steady evolution of military art, and will create 'a qualitative break between the means of past and future war'. The key to this exercise is the search for 'law-governed patterns' (*zakonomernosti*), which identifies what is relevant from past combat experience and what is in the process of being radically transformed.

Gareev emphasizes two profound changes that are making for a radical break in military art: The first is the transformation of the international system resulting from the end of the Cold War, and the associated political, social, and economic realignments that have occurred. The second is the revolution in military affairs based on the appearance of advanced, high-accuracy conventional weapons, electronic warfare, and information warfare. Taken together, these developments call for a 'total analysis' of 'socio-political and economic factors, the development of armaments on the basis of the latest technological innovations, changes in the human material associated with the relationship of society to war, and to the methods of raising the armed forces'. Gareev reasserts the relevance of Clausewitz to post-Cold War security issues, saying that states, not weapons, make war. 'Therefore it is more correct to say that under contemporary conditions war – nuclear and conventional – should not be a means and continuation of politics.'

Concerning changes in the international system, Gareev notes the sharp decline of ideology as a source of conflict, but warns that 'new, quite sharp socio-political, economic, territorial, and ethno-national contradictions between various states and regions' have emerged as dangers and threats. Gareev criticizes the Russian government for making one-sided concessions and failing to act in a manner that would protect and support Russian national interests. Defense on the cheap is an invitation to loss of sovereignty, independence and territorial integrity.

Gareev begins by addressing the need to emancipate military art from nuclear weapons, which he describes as weapons that are impossible to use. He notes that for the foreseeable future nuclear weapons will remain 'a cheap and reliable means of deterring aggression'. He sees serious political liabilities in the provisions of Russia's current military doctrine which reject a declaratory policy of no-first-use. In a political crisis, an assured second-strike capability reduces the risk of either side gambling on preemption. While Gareev sees a low probability of a conventional world war over the next decade and a half, he does note the trend among states to use subversive actions and local wars to achieve political objectives, and warns that such conflicts could evolve into large-scale armed conflicts. In this context, strategy will be dominated by B.H. Liddell Hart's 'indirect approach', to achieve limited ends with limited means. Conflict prevention and localization take on increased significance in this environment and involve 'economic sanctions, naval, air, and ground blockade of communications, demonstrations of force, the deployment of peacekeeping forces for the separation of the sides and other means of action'. Failure to prevent conflicts or bring about their isolation will lead to escalation, including offensive actions aimed at the military power of the offending party. The model here is the Gulf War, with its initial independent air operation followed by a ground offensive to take and control key terrain and crucial military objectives. Gareev sees a major challenge before military art in addressing those problems that arise in conflicts between more technologically-advanced and less-advanced states.

On the revolution in military affairs, Gareev foresees profound shifts in strategy, operational art and tactics as a result of technological innovations which will shift the balance among fire, strike and maneuver; because advanced precision weapons, electronic warfare, and modern troop control systems will allow simultaneous fire and strike maneuvers throughout the depths of an enemy's dispositions, in a theater Gareev calls 'sea–air–land operations'. Such an operation could lead to the disruption and destruction of important enemy combat groupings. Such combat actions would link together battles on the front, flanks and rear of an opponent in a non-linear battlefield of great extent, without stable front lines, and involve significant maneuvers in depth. The impact of these changes would be felt most in the initial period of war, and would place great stress on systems of troop control, requiring greater reliance on automated control systems by commanders and staff. Gareev raises the issue of seeking international arms control agreements to deal with 'strategic conventional weapons', by which he means advanced precision systems that could be used for strategic decapitation of national command and control capabilities in the initial period of a future war, thereby forcing nuclear states without adequate means to defend against such strikes to rely on strategic preemption in a crisis period.

Gareev recognizes that Russia's military is in a complex process of reform and restructuring following the end of the Cold War and the disintegration of the USSR. And that the socio-economic and political transformation of Russia will either guide the creation of a post-Soviet military, or be held captive by the inherited military structures that are too large to be reformed and too ossified to undertake reform themselves. Those who focus on the present might dismiss this volume on the simple basis that Russia, with all its internal problems, and the armed forces' tragic performance in Chechnya, is in no position to address the issues or carry out the economic transformations to be an effective military competitor in the first decades of the next century. Based on historical experience, such a view would be very short-sighted. Defense analysts, statesmen, and soldiers will find much food for thought in this volume. Part of the credentials of this forecast belong to the author himself, a soldier with more than fifty years of active service.

THE MAN

General Gareev, one of the leading military theorists of the Soviet Union and the Russian Federation, brings to his chosen topic a wealth of practical experience as a soldier, trainer, military historian and forecaster. Born in

1923 in Chelyabinsk, Gareev is one of the last of the famous 'cohort of 1923', the young men born in that year who went off to fight in 1941, survived the Great Patriotic War, and went on to lead the Soviet armed forces during the Cold War.[1] His generation was shaped by the Stalinist revolution from above that began in the late 1920s with industrialization and collectivization. His family left Chelyabinsk in 1932 in the midst of famine and traveled for six months to Tashkent, where his two older brothers had reported better conditions. On the trip the family lived in shanty towns at rail junctions, barely making do. When the family arrived in Tashkent they found hard times there too.

In 1935 Gareev's father placed him in a military school in Tashkent to reduce the economic burden on his large family. The young Gareev attended a cavalry school, where he played in the band, before he was accepted into officers school in 1939. He completed officers school in the spring of 1941 in time to join the fighting of winter 1941 before Moscow.[2] General Gareev's prognoses on the changing trends in armed struggle are, therefore, informed by his own extensive experiences as a soldier, and reach back almost six decades to Soviet Central Asia and the cavalry that fought the Basmachi. These include his experiences as a combat soldier with the Western, Third Belorussian, and Far Eastern Fronts during the Great Patriotic War. Gareev experienced the transformation of the Red Army from an infantry army that fought before Moscow, into the tank and mechanized forces that liberated Soviet territory and carried the war into Eastern Europe and Germany. He was in the Far East for the lightning operation of August 1945 that defeated Japan's Kwantung Army. Like most of his cohort, he was wounded several times in the course of the war.

In the post-war years Gareev held positions of increasing importance in the Soviet armed forces: various command and staff positions in the Far Eastern, Belorussian and Ural military districts, and with the Soviet Forces in Germany. He was an active participant in the nuclear revolution as a rational instrument of policy, with its expanding arsenals of destructive power and narrowing utility. He served as chief of staff to the main Soviet military adviser in Egypt in the early 1970s, including during the air war of attrition over the Suez Canal. In early February 1989 Gareev assumed the post of chief Soviet military adviser to the president of Afghanistan. He held that post until 1990. With a small operational staff, Gareev advised the Kabul government's forces how to conduct local war after the departure of the Soviet 40th Army.[3] In both Egypt and Afghanistan Gareev displayed an attention to training that gave combat stability to unstable armies. Within the Soviet general staff Gareev was noted for his work on training and military science. He is the author of *Tactical Exercises and Maneuvers* and *Combined Arms Exercises*. Western audiences are familiar with his study of the military theorist M.V. Frunze which has been translated into English. During Marshal Ogarkov's tenure as chief of the Soviet general staff, General Gareev played a leading role in the development of the operational-maneuver group concept, for the conduct of deep, air-mechanized operations.

The Russian military, and especially the general staff, inherited from the Soviet military the firm belief that revolutionary changes in the nature of warfare are afoot, and that these innovations demand significant changes in the military–technical side of military doctrine and in the approach to military force structure. Beginning with Marshal Ogarkov's tenure as chief of the general staff, Soviet military analysts began to speak of a revolution in military affairs. They associated it with a new generation of nuclear weapons and the appearance of advanced, high-precision conventional weapons. General Gareev, then deputy chief of the Soviet general staff and chief of the directorate for military science, described the RMA (revolution in military affairs) in the following terms:

> Now we can speak about a turning point in the development of military science and military art. In general, a new qualitative leap in the development of military affairs, connected with the modernization of nuclear weapons and especially the appearance of new types of conventional weapons, is ripening. In connection with this [process] there has arisen the need to rethink the basic military-political and operational-strategic problems of the defense of the socialist Fatherland.[4]

Gareev's last active duty assignment was as deputy chief of the general staff and head of the directorate of military science. General Gareev is the organizer and first elected president of the recently founded Academy of Military Sciences. He continues to write on military history and military theory and is presently completing his study of Marshal G.K. Zhukov's operational art. His military experience spans the era from cavalry to information warfare. He has confronted the challenge of change, which has so often recast the art of war in this century.

THE ISSUE

Gareev's book should be seen against the backdrop of an intense debate over the revolution in military affairs (RMA) and the impact of information warfare on the conduct of war. The Gulf War initiated an intense debate in the United States, the Soviet Union/Russia and elsewhere over the very existence of a 'revolution in military affairs'. Since the end of the Cold War the RMA's significance for US defense planning has become part of the ongoing conflict over military downsizing, funding of current operations, and maintenance of the US's technological initiative into the next century. These exchanges have become increasingly intense. The two positions, pitting advocates against doubting Thomases, contrast a revolutionary interpretation as opposed to an evolutionary one. In the former case, the Gulf War represents the harbinger of radical changes, transforming warfare as profoundly as have mechanization and the introduction of nuclear weapons.[5]

This revolutionary interpretation sees the RMA as a transformer of combat through the appearance of advanced, high-accuracy precision strike weapons, advanced systems of C3I [command, control, communication and intelligence], electronic warfare, and computer simulation. Quality forces will be those equipped, organized, and trained to make use of advantages in information, penetration, and precision against an opposing force. Describing the operational environment of land warfare in the twenty-first century, General Gordon Sullivan, former chief of staff of the US Army, and his co-author, LTC James M. Dubik, spoke of five trends:

> Greater lethality and dispersion.
> Increased volume and precision of fire.
> Better integrative technology leading to increased efficiency and effectiveness.
> Increasing ability of smaller units to create decisive results.
> Greater invisibility and increased detectability.[6]

These trends are reshaping warfare towards a joint endeavor in which synergy is achieved through simultaneity.

Forces of these kinds will be able to achieve 'a qualitatively different way of fighting – the ability not only to strike the enemy deep, but to see the enemy deep in real time'.[7] General Gordon Sullivan, former chief of staff of the US Army, and his co-author, LTC Anthony Coroalles, have pointed out: 'With this capability, commanders can now blend previously separate and discrete operations into a single and seamless whole'.[8] They identified this as the strategic-theater operation, and pointed to Operation Urgent Fury in Panama in December 1989, and Desert Storm in the Persian Gulf in early 1991 as examples of modern operations by joint forces using 'simultaneous employment of overwhelming combat power throughout the breadth and depth of the operational area to paralyze the enemy'.[9] By linking surprise, simultaneity 'in time, over time and throughout the levels of war' with an assault on the enemy's decision cycle, the Allies in the Gulf War 'achieved the effect of simultaneity over the Iraqis at *all levels of war*'.[10]

Yet the Gulf War was only a 'glimpse' of those capabilities that are now, and will continue, to reshape warfare in the information age. That this glimpse coincides with the end of the Cold War, a general reduction in forces, and a radical recasting of the international environment makes foresight particularly difficult. In this regard the examination of the historical record on other revolutions in military affairs is both interesting and rewarding. There are two distinct positions in such revolutions. One position sees such revolutions as infrequent events, resulting more from profound changes in societies which generate military power. This minimalist view would accept the term revolutionary to deal with shifts in the dominant organizing power behind the military instrument: animal, mechanical, industrial, nuclear, and informational. Each revolution in this scheme of things turns out to have been a complex, protracted, and ultimately profound process, leading to the transformation of warfare and military art. From this viewpoint most of the changes in military art are, in fact, evolutionary. The other, maximalist position, sees modern industrial war throwing up the challenge of persistent change, fed by the transformations in societies' economic orders and social organizations. Permanent revolution in this case becomes the very model for change. Both models explicitly recognize that the nuclear revolution, which they both accept, has worked a profound change in the relevance of war as a rational instrument of politics. By creating the means for absolute war in such weapons of mass destruction, the states possessing such weapons were forced to contemplate the limitation of war as an instrument of policy. The very possession of such arsenals imposed a deterrence posture and an acute need to limit the escalation of conflicts into general wars. The situation did not lead to the end of armed conflict, but it did drive such conflicts in new directions and gave such conflicts new contours. It is in this context that the Gulf War takes on new importance as a harbinger of things to come.

THE INFORMATION RMA AND THE HISTORICAL RECORD

Twentieth-century warfare, up until the Gulf, was dominated by changes wrought by the industrial revolution. With regard to military art, industrialization had two immediate and profound consequences. First, it made possible the really massive armies of the late nineteenth century. While the French Revolution had promoted the concept of the nation-in-arms through the *levee en masse*, industrialization created the wealth that states could draw upon to mobilize, train, and arm their citizens/subjects. Second, industrialization provided the material to sustain massive military efforts, reducing the decisiveness of major battles and transforming strategy from the art of concentration of combat power at a single point, into a linear battle in which each side sought to exploit an open flank until none existed. Jan Bloch, born Ivan Bliokh, the Warsaw banker and railroad entrepreneur, saw this trend almost two decades before World War I began and noted the grave risk that such a defense-dominated, industrial war posed for the societies fighting it. The front could be maintained, but at the risk of revolutionary upheaval in the rear.[11] At that juncture, the core issue of mass industrial war became the ability to strike the opponent deep by negating his linear defense before stalemate brought about socio-economic and political collapse. What happened in the Gulf War is analogous to the use of new technology and tactical techniques in the final year of World War I to overcome that linear stalemate of trench warfare. New techniques and weapons restored tactical maneuver, and opened the prospect of breakthroughs and operational success in that theater. The German Army relied on tactical innovation, associated with the use of shock troops. The Allies, following the British lead, embraced technological innovation through the introduction of the tank and its massed employment in support of infantry assaults. At the end of World War I the full implications of these innovations remained unclear. Theory outran existing capabilities, and military innovation in a time of reduced funding and low threat perception proved slow. Only two decades later did a synthesis of these innovations emerge as *Blitzkrieg* in Nazi Germany. Here, the massed employment of panzer divisions, embodying the tactical integration of tanks, radio communications, and aviation attack, restored maneuver and defeated opponents who fought a positional war. By the final stages of World War II the Soviet Army had achieved the organizational maturity and material support to put deep operations into practice. The mature Soviet approach to mechanized warfare had an operational focus. Tank and mechanized formations, airborne forces, and air armies conducted deep-battle, and deep successive operations to achieve the destruction of the enemy force throughout the depths of its deployment and across vast continental theaters. These capabilities had a solid pre-war foundation in Soviet studies of operational art in the inter-war years, in preparation for a general war in eastern Europe, and in the draconian industrialization carried out during Stalin's 'revolution from above'.[12]

A similar argument can be made for a revolution in military affairs with regard to the development of carrier aviation and amphibious warfare capabilities by the United States and Japan in the inter-war years. Once again Britain had the lead in both fields at the end of World War I, and for a host of geo-strategic, fiscal, organizational, and doctrinal reasons abdicated leadership in these areas to the United State and Japan, whose naval élites in preparing for an eventual confrontation in the Pacific could appreciate the radical implications of these instruments for that theater. Here again wartime experience brought a maturation of the concepts, techniques, organizations, and instruments to bring the revolution to full maturity, as embodied by the carrier task forces and invasion fleets of 1944–45.

CRITICS AND PRECURSORS

In all previous cases, practitioners have adapted their RMA to the requirements of a particular theater, and have depended upon a theater infrastructure to sustain and support its successful application. Thus the situation after the Gulf War, in this interpretation, was a challenge to innovation. The revolution in military affairs being the product of the information society only now being born, the very pace and scope of change within that society provide the form and substance to an ongoing, but still undefined, military revolution.[13] Critics dismiss this interpretation of the revolution in military affairs. They portray it as a slick device the US military is using to retain a disproportionate share of reduced federal spending in the aftermath of the Cold War.[14] Changes are happening, but are they evolutionary or revolutionary?

Critics disagree on the interpretation of the Gulf War and its significance. But even before that conflict, a debate over a new revolution in military affairs was under way. Some analysts drew on military history to identify a qualitative change in the evolution of military art, and began forecasting a significant shift in the nature of war. The essence of modern war as a social phenomenon changed radically over the last century. The twentieth century began in August 1914 with the arrival of mass industrial war, and total war. The instruments of war then outgrew their political utility until a major military confrontation between the two militarized blocs became unthinkable. Military power could be, and was, used in local conflicts. In their strategic forms nuclear arsenals deterred the use of other forms of military power to resolve the core contradictions between the west and the Soviet bloc in Europe. Conventional weapons also evolved in response to new technologies, and began to reshape military art. In the mid-1980s the late Brigadier Richard Simpkin, drawing heavily on Soviet military theory, made a compelling case for the need for military theory to lead technological innovation. He pointed to the further adaptation of deep battle to the emerging concepts of operational maneuver and air mobility associated with the Soviet operational-maneuver group (OMG).[15] With the development of deep-strike precision systems associated with NATO's follow-on forces attack (FOFA) concept, new discussions over a shift in the nature of fire and maneuver emerged.[16] Writing on the eve of the Gulf War, Christopher Bellamy called attention to the changes in technology that are reshaping warfare.

> Advanced military forces are dependent on computers, radio and other communications, and satellites, for reconnaissance, navigation, and communications. Attacking the enemy's 'brain and stomach' need no longer depend on tanks racing round a flank, or aircraft pounding headquarters and industrial centres from above. The enemy's brain and nerve system can be seared and paralyzed by jamming, and various types of electromagnetic weapons. Electronic warfare, and other 'soft kill' weapons are likely to usurp the position envisaged for tanks and aircraft in much of the 1930s military theory. Low-frequency weapons and application of bioelectronics may severely reduce the effectiveness and alertness of enemy forces, commanders, and political leaders.[17]

Bellamy asserted that large-scale land warfare among major powers had evolved to a dead end by the finish of the Cold War. Mass warfare will give way to local wars, fought by forces configured for such special operations. Political changes, especially the emergence of a multipolar world, will push military forces toward greater professionalism and new technologies.

ARMED CONFLICT AND THE INFORMATION AGE

The core issue in this debate is the relationship between war and society. The unrelenting pace of technological innovation, fundamental shifts in the subject and organization of production, the vast recasting of institutions, and rapid shifts in social values raise the prospect of a self-organizing, adaptive society in a state of becoming, making the current era truly revolutionary. Military institutions must adapt to this challenge within their own societies and prepare for the emergence of new and potential conflicts within and among states and non-states. Ethno-national conflicts, ecological threats, demographic catastrophes (mass internal migrations, famine, epidemics, and the flight of refugees affecting entire states and regions) and large-scale social pathologies with transnational dimensions (drug-trafficking, organized crime as a social movement, and terrorism) form the new matrix of threats in an unstable world. How military institutions might adapt to these challenges is at the very core of the discussion of the revolution in military affairs. It is here that technological and social change meet.

As popularized by Alvin and Heidi Toffler, the post-industrial information society has produced the potential for 'third wave' warfare, which transcends industrial war in the same manner that the former negated agrarian warfare.[18] Recognizing the persistence of earlier forms of warfare and their intermingling in conflicts fought by societies at different stages of development, the Tofflers assert that revolutions in military affairs are quite rare and reflect fundamental shifts in the very organizing principles of societies. It is not to be confused with a mere evolution in military art. Military institutions, which seek to adapt and evolve gradually, will find their response inadequate, untimely, and possibly fatal. The Tofflers warn:

> A true revolution goes beyond that to change the game itself, including its rules, its equipment, the size and organization of the 'team', their training, doctrine, tactics, and just about everything else. It does this not in one 'team' but in many simultaneously. Even more important, it changes the relationship of the game to society itself.[19]

Racing toward a post-industrial society, the advanced nations of the world are contesting to see who will reap the greatest benefits from this transformation. 'The global competitive race will be won by the countries that complete their Third Wave transformation with the least amount of domestic dislocation and unrest.'[20] In military terms the United States, according to this interpretation, has already made the first step to third wave warfare with the development of air/land battle and advanced, high-precision deep-strike weapons, and has applied them successfully in the Persian Gulf War against a regional, second-wave industrial military.[21] Their discussion of third wave warfare asserts the need to create new instruments to prevent small conflicts from engendering major wars. Thus, the war and anti-war of their title involves the adaptation of new ways of 'understanding ... the revolutionary new linkage between knowledge, wealth and war'.[22] Darwinists in their view of this race to the swift in creating an information society, the Tofflers see serious global problems arising from the very process of creating this new order. As Alvin Toffler asserted in a recent interview, this future will have its own sources of conflict layered on top of existing sources, and its own terror weapons.

> Also, there will be massive dislocations. Just like at the time of the Industrial Revolution. And this current change is even bigger, moving faster and covering more of the planet. So there will be a lot of social upheaval. There are terrifying pieces in the future. Race-specific weaponry. You can zero in on ethnically linked genetic characteristics and target those who carry them. This is *genetic* warfare, a modern version of giving the Indians infected blankets. Terrifying. Absolutely.[23]

Accounts of information warfare in the popular press paint an even more dramatic image of future warfare in which the command of 'cyberspace' becomes the precondition of all other successful combat actions, and where there are covert strikes against enemy control systems, ranging from air defense and troop control to energy, transport, and national communications.[24] At the same time, such accounts stress the vulnerability of advanced societies to such strikes themselves, but without noting the high degree of interdependence and interpenetration of such societies in the global market. Moreover, initial leadership in any advanced field of technology, even in those upon which the information age is based, does not necessarily translate into long-term military superiority, if others can find the means to lever their initial backwardness to an advantage. Such was the case with mechanization for Britain, and with nuclear weapons for the United States. And such can be the case for the future, as was pointed out in an article devoted to future war in the information age.[25]

THE CHALLENGE OF CHANGE

While the Tofflers have been dismissed by some critics as popularizers selling futurist snake oil to the gullible, their wide-ranging forecasts and persistent emphasis on the challenge of change to human society call attention to the need to grasp the interconnections among trends in diverse and seemingly unconnected fields of endeavor. As consultants, the Tofflers have made a career out of advising practical leaders in business and government of just such trends. Their recent attention to the RMA in this regard has much in common with that of Russian military forecasters, who have been trying to fathom its 'law-governed patterns' (*zakonomernosti*)for almost two decades.[26] The problem for soldiers and military analysts alike in the United States, Russia and elsewhere is to deal with the challenge of change itself. General Gordon R. Sullivan and LTC Anthony M. Coroalles have focused on this aspect of the challenge of change, specifically, its accelerating pace, qualitative character, and complexity. By invoking a metaphor from the American Civil War, 'seeing the elephant', they have sought to draw attention to the 'complexity, ambiguity, and uncertainty of tomorrow's battlefield'.[27] On the military–technical side of the RMA, the challenge of change is clearly and explicitly being addressed in the US armed forces by an attempt to adapt military art (doctrine in US military usage) to information warfare.

> Simply stated, Information Age technology can fundamentally change the way that we prepare for and conduct military operations

> of all types. We believe that information technologies can enable the Army to achieve quantum improvements in speed and precision – enhancing tempo, lethality, and survivability to new levels. By being able to integrate operations more effectively, we will achieve unprecedented and overwhelming degree of simultaneity in the Joint Force. Our goal is to use the power of information to leverage the power of our people, so as to change how we develop, sustain, and constantly improve the Army.[28]

The authors go on to outline an approach to making change happen that stresses understanding the environment, articulating vision, building a consensus for change, creating a learning organization, fostering innovation and growth, establishing a convincing demonstration of value, and implementing change by providing the resource to carry it through.[29] They discuss those programs now under way to implement change, and stress the impact of down-sizing, a new strategic environment, redeployment to the continental United States and the emergence of new missions on this process of military-technological change. Ultimately, success in meeting the challenge of change also depends on continuity, 'the self-ordering nature of our shared values' built around a military professionalism based upon 'faithful service to the nation'.[30]

THE RUSSIAN RESPONSE

The revolution in military affairs has deep roots in the changes that are rapidly recasting advanced industrial societies into information societies. The relationship of military institutions to the state and society will experience as profound a change as that which transpired in the shift from agrarian to industrial society. Moreover, the global nature of the information age with its high degree of interdependence and interpenetration will change the international system and recast the very meaning of threat. The consequences of uneven development, which served as so great a source of conflict among various states at different stages of industrial development, and between the industrial powers and the agrarian-colonial regions, can only become more complex in the case of information societies and those at lower stages of development. Successful adaptation of national power in all its forms to information warfare will depend in good measure on the impact of these developments on the nation and the international environment.

The end of the Cold War still holds out the prospect that nations may find good reasons to cooperate in employing these new means to prevent conflicts, or, if not prevented, to limit their escalation, and assist in their resolution. But this will require that statesmen likewise 'see the elephant'. In the realm of politics and political economy, the challenge of change is even more evident and the need for foresight to guide military-political development all the more imperative. Democratic and open societies have the greatest chance of finding common ground to address the future and to seek the means of enhancing peace and stability, even as they meet the challenge of change. This is particularly important as former Cold War adversaries seek to find common grounds. This will not be an easy task in so far as one side sees itself as victor and the other as vanquished. Yet, the insights of the loser in analyzing his defeat, however bitter in tone, may often offer the more interesting perspective on past, present, and future trends.

In a recent interview on military–technical policy, General Gareev offered a retrospective analysis of the failings of the Soviet leadership in managing its military competition with the west. In the absence of a rational policy to keep the burden of defense within tolerable limits, the Soviet Union found itself consumed by the ever increasing costs of global competition. Gareev writes:

> In the course of the last twenty-five years every time a new Minister of Defense or Chief of the General Staff was appointed I immediately wrote a report about the basic necessity of the transformations in our armed forces, demonstrated that they were artfully pulling us in the direction of arms races, in which we should not have participated and that we should have sought some way of avoiding. We should not fall behind in military–technical relations and maintain strategic parity. But there was no need to embark on everything that the Americans had done.[31]

Gareev goes on to point to SDI (Strategic Defense Initiative) as one such program, which the US leadership used to bankrupt the USSR, and identifies the Soviet military-industrial complex as a willing co-conspirator in that process. Gareev states: 'According to a whole series of positions, especially regarding electronics and computers we were always behind and to compete with the Americans in the realization of programs like SDI was simply impossible.'[32] In general, Gareev suggests that the USSR collapsed from strategic overreach associated with the global competition with the US.[33] Gareev goes on to assert that under Marshal Ogarkov the general staff had proposed significant cuts in defense spending in a number of areas: civil defense, aircraft carriers and strategic defense; and in reducing the number of branches of the armed forces including abolishing PVO [Air

Defense Forces] as a separate branch. According to Gareev, Ogarkov also called for a reduction in Soviet military assistance to the Third World, which he categorized as 'ideological expenditures', unconnected to specific national interests. Gareev suggests that one of the reasons the Soviet Union failed to engage in effective military foresight was the fact that there were no independent think-tanks engaged in assessing these trends. Gareev points to the independent, non-governmental Academy of Military Sciences, which he helped to found, as such an agency.[34]

With regard to his advice on the development of the Russian military-industrial complex to meet the needs of the revolution in military affairs, Gareev points to micro-electronics as the key area. Not only was the USSR clearly behind in this arena, but micro-electronics are the foundation of the revolution. What is the perspective for weapons development? 'The main [direction] is not individual types of weapons, but their complexes, which in informational relations integrate: means of reconnaissance, identification, observation, and electronic warfare and all that is connected with it.'[35] The key to victory in modern armed conflict becomes the ability to find the enemy before he finds you, and to employ weapons systems of high accuracy and lethality.

Gareev bows in the direction of his American opponents during the Cold War, pointing to their pragmatism. He has no doubt that Soviet military science was more developed and made a greater contribution to military art and theory. However, he notes that Russian and Soviet politicians lacked the ability to create optimal conditions for the application of military power. Looking at the experience of the Gulf War, Gareev credits Amercian political leadership with the ability to create such optimal conditions for the employment of the armed forces in the service of the national interest. Indeed, he accuses the US of fomenting the conflict in the first place.[36] He concludes that Russia could not have mounted so effective a campaign to prepare for war politically, or brought so much high technology to bear. He notes that Russian military planning for such an operation would have been distinct but not inferior. 'But that is one side of the matter.'[37]

Turning to the evolution of strategy in the late twentieth century, Gareev sees the combination of a new international era and the RMA leading to a fundamental change. 'Now a new era has begun, when it is especially important to understand that indirect actions give no less effect. The highest manifestation of such actions is the tremendous work done toward the subversion of the USSR.' He includes in this the west's efforts to undermine the Soviet economy, prevent the import of advanced technology, draw the USSR into the arms race and global confrontation, and the weakening of allies, such as Cuba, North Korea, and others, and finally internal subversion, including the erosion of its ideology. 'The sum total of all of that led to the result which Hitler had not been able to achieve during four years of war with the USSR.'[38]

It is in this context that the English translation of General Makhmut Akhmetovich Gareev's book on future armed conflicts should be of interest to western military specialists. The book appeared in its Russian edition *If War Came Tomorrow?* in 1995.[39] The Russian title invoked the pre-war years, when preparations for the inevitable war with the encircling capitalist powers was seen both as imminent and as the ultimate test for the Soviet state and society. The Russian edition was greeted with positive reviews in leading military-scientific publications. In 1996 a German edition of the book appeared under the title *The Contours of Armed Conflict in the Future*, with its forecasting claims reduced to the next 10–15 years.[40] The Federal Institute for Eastern and International Studies in Cologne sponsored its publication. The present translation should contribute to a dialogue among military specialists and statesmen, who must be concerned with the trends that will shape armed conflict in the future. Having been in two projects which brought classic Soviet works of military theory to the attention of western specialists – Svechin's *Strategy* and Triandafillov's *The Nature of Operations of Contemporary Armies* – I am certain that readers will find much of interest here.[41]

NOTES

1. The 'cohort of 1923' which included such well-known Soviet military officers as Marshal Sergei Akhromeev and General Varennikov, fought the war from beginning to end, experiencing the initial defeats, the struggle for the initiative and the great offensives of the last two years of the war. These men joined the Communist Party while in wartime service and had little doubt about the justice of the cause they served during the war or in the Cold War that followed.
2. Conversation with General Gareev in Moscow, Russia, on 26 July 1996.
3. M.A. Gareev, *Moya poslednyaya voyna (Afganistan bez sovetskikh voysk)* (Moscow: INSAN, 1996) pp. 5–7.
4. M.A. Gareev, *M.V. Frunze – voyennyy teoretik* (Moscow: Voyenizdat, 1985) p. 438.
5. Andrew Krepinevich, 'Cavalry to Computer', *The National Interest*, 37 (Fall 1994) pp. 30–42.
6. Gordon R. Sullivan and James M. Dubik, *Land Warfare in the 21st Century* (Carlisle Barracks, PA: Strategic Studies Institute, 1993).
7. Gordon R. Sullivan and Anthony M. Coroalles, *The Army in the Information Age* (Carlisle Barracks, PA: Strategic Studies Institute, 1995) pp. 11–12.
8. Ibid., p. 12.
9. Ibid.
10. Ibid., pp. 14–15.
11. I.S. Blokh, *Budushchaya voina v tekhnicheskom, ekonomicheskom i politicheskom otnosheniyakh* (St Petersburg: Tipografiya I. A. Efrona, 1898).
12. Jacob W. Kipp, 'Mass and Maneuver and the Origins of Soviet Operational Art' in Karl Reddel (ed.), *Transformation in Russian and Soviet Military History: Proceedings of the Twelfth Military History Symposium, USAF Academy, 1986* (Washington, DC: United States

Air Force Office of Air Force History, 1990) pp. 87–116; and 'Soviet Military Doctrine and the Origins of Operational Art, 1917–1936', in: Philip S. Gillette and Willard C. Frank, Jr. (eds), *Soviet Military Doctrine* (Lexington Books,1992).

13. 'The Software Revolution: A Survey of Defense Technology', *The Economist* [insert] (10–16 June 1995) pp. 5–11.
14. A.J. Bacevich, 'Preserving the Well-Bred Horse', *The National Interest*, 37 (Fall 1994) pp. 43–9.
15. Richard Simpkin, *Race to the Swift: Thoughts on Twenty-First Century Warfare* (London: Brassey's Pergamon Group, 1985); and *Deep Battle: The Brainchild of Marshal Tukhachevskii* (London: Brassey's Defence Publications, 1987) p. ix.
16. Jacob W. Kipp, 'Conventional Force Modernization and the Asymmetries of Military Doctrine: Historical Reflections on Air/Land Battle and the Operational Maneuver Group', in: Stockholm International Peace Research Institute, *The Uncertain Course: New Weapons, Strategies and Mind Sets* edited by Carl G. Jacobsen, (Oxford: Oxford University Press, 1987) pp. 137–66.
17. Christopher Bellamy, *The Evolution of Modern Land Warfare* (London: Routledge, 1990) pp. 242–3. For the views of the author on this topic see: Jacob W. Kipp, 'The Evolution of Soviet Operational Art: The Significance of "Strategic Defense" and "Premeditated Defense" in the Conduct of Theatre–Strategic Operations', *The Journal of Soviet Military Studies* 4, No. 4 (December 1991) pp. 629–48; and 'Soviet Military Foresight and Forecasting in an Era of Restructuring', in: Derek Leebaert and Timothy Dickinson (eds), *Soviet Strategy and New Military Thinking* (New York: Cambridge University Press, 1992) pp. 248–75.
18. Alvin and Heidi Toffler, *War and Anti-War: Survival at the Dawn of the 21st Century* (Boston: Little, Brown and Company, 1993) p. 29.
19. Ibid., p. 29.
20. Ibid., p. 25.
21. Ibid., pp. 64–80.
22. Ibid., p. 252.
23. Claudia Dreifus, 'Present Shock', *The New York Times Magazine* (11 June 1995) p. 48.
24. 'Onward Cyber Soldiers', *Time* (21 August 1995) pp. 37–47.
25. Thomas E. Ricks, 'Warning Shot: How Wars are Fought Will Change Radically', *The Wall Street Journal* (15 July 1994) pp. 1, 5.
26. For several recent Russian works addressing future warfare see: Eduard Shevelev, 'Through the Creation of a Single Contemporary Military Science and its Worldwide Acceptance', speech delivered at National Defense University, 15 March 1991; V.I. Slipchenko and E.G. Shevelev, 'Metodologiya reformy Vooruzhennykh Sil', *Voyennaya mysl'*, Nos. 11–12 (Nov.–Dec. 1991); V. Larionov, 'Possible Consequences of the Intellectualization of Weaponry', paper prepared for 16th USAF Academy Symposium on Military History (Colorado Springs: USAF Academy, 1995; V. Aleksin and E. Shevelev, 'O reformirovanii nashego flota', *Morskoy sbornik*, No. 3 (March 1995) pp. 9–11. See also: Jacob W. Kipp, 'Russian Military Forecasting and the Revolution in Military Affairs: A Case of the Oracle of Delphi or Cassandra?' A paper delivered at the Military Operations Research Society Symposium, Annapolis, MD: US Naval Academy, in June 1995 and published in *The Journal of Slavic Military Studies* in 1996.
27. Gordon R. Sullivan and Anthony M. Coroalles, *Seeing the Elephant: Leading America's Army into the Twenty-First Century* (Cambridge, MA: Institute for Foreign Policy Analysis, 1995) pp. 1–2.
28. Ibid., p. 13.
29. Ibid., p. 23.
30. Ibid., p. 47.
31. M.A. Gareev, 'Voyenno–tekhnicheskaya politika: Retrospektivnyy analiz', *Problemy prognozirovaniya*, No. 3 (March 1996), p. 163.
32. Ibid., p. 166.
33. Ibid., p. 164.
34. Ibid., p. 165.

35. Ibid., p. 167.
36. Ibid., p. 169.
37. Ibid.
38. Ibid., pp. 169–70.
39. M.A. Gareev, *Esli zavtra voyna?... (Chto izmenitsya v kharaktere vooruzhennoy bor'by v blizhayshie 20-25 let)* (Moscow: Vladar, 1995).
40. Machmut Achmetowitch Garejew, *Konturen des bewaffneten Kampfes der Zukunft: Ein Ausblick auf das Militaerwesen in den naechsten 10 bis 15 Jahren in: Schriftenreihe des Bundesinstituts fuer ostwissenschaftliche und internationale Studien, Koeln, Band 28* (Baden-Baden: Nomos Verlagsgesellschaft, 1996).
41. Jacob W. Kipp, 'General-Major A.A. Svechin and Modern War: Military History and Military Theory', Introductory essay for: Kent Lee (ed.), A.A. Svechin, *Strategy* (Minneapolis: East View Publications, 1992); and Jacob W. Kipp (ed.), *V.K. Triandafillov's 'The Nature of the Operations of Contemporary Armies'* (London: Frank Cass, 1994).

CHAPTER 1

POLITICAL AND MILITARY–TECHNICAL FACTORS AND ARMED STRUGGLE

Future Wars: Sources and Causes

There are growing tendencies in the world to suppress war and make the unlimited use of armed force impossible. Their reaches are ramifying, causing a need to find new military–political and strategic conceptions.

1 WAR AND POLITICS

Historically, wars had economic roots and underlying political motives. However, their actual objectives and causes, as a rule, were disguised as religious, ideological, and other motives. Therefore, it was not until the nineteenth century that the connection between war and politics became apparent, when, in addition to armies, great masses of the population were drawn into war.

Since Clausewitz it has been acknowledged that war was a continuation of politics by other and violent means. But since the advent of nuclear weapons another standpoint has prevailed that war no longer is a continuation of politics. There are two separate issues involved here: one is the question of expediency and admissibility of war in the modern world; the other is the question of whether new technologies can alter the socio-political essence of war.

The answer to the first question can and must be formulated in no uncertain terms: all wars must be banned from the life of the human race. Particularly inadmissible is nuclear war, fraught with catastrophic consequences for the participants and for other countries. Yet despite the obvious correctness of this postulate, and contrary to the best intentions, it is not always possible to resolve contradictions by political means and thus various forms of military violence continue to persist, even though we live in the nuclear age.

To this day, no foolproof guarantee exists that numerous local wars and conflicts will not evolve into large-scale warfare.

It is easy to follow trendy conceptions, avoiding the essence of the issue, but actual phenomena exist regardless of our ignorance or criticism. One needs to understand their essence in order to respond to them correctly. What do all these incessant wars and conflicts mean, and what is their origin if politics are believed to have already excluded armed methods from their agenda?

Academician G. Arbatov writes that, 'even in local conflicts it [military force] cannot be regarded as an instrument to achieve rational political goals – at least in conflicts larger in scale than Grenada, Panama and the Falklands'.[1] Yes, it is true that at present it is not reasonable to connect rational policy with military force. But rational policy is often accompanied by irrational policy, which produces phenomena typical of it. Additionally, judgment of whether a policy is rational and reasonable, or not, is of a partial, circumstantial and debatable nature.

Some states really want peace. In others, ruling circles declare that there are issues more important than peace, or that there are worse things than war. All of them are confident of the rationality of their policies. And what should be done if war is imposed on a country, or its national sovereignty is endangered? Life poses such questions and they cannot be left unanswered. In Arbatov's opinion, the Falklands War can still be viewed to some extent as the means of a rational policy, but what about the Iran–Iraq War; the Gulf War of 1991; or the current hostilities in the Caucasus and many other numerous armed conflicts? Yet the same Falklands War could have evolved into a much larger conflict, and that would not have altered its true nature defined by certain political goals of the warring sides. If this is not so, an explanation must be given of why and to what end wars like those continue to break out. So far, there has not been any evidence that any of them had other than political causes.

The essence of war (once it has been unleashed) cannot be changed by the use of nuclear weapons as well. M. Gorbachev once said that after Hiroshima and Nagasaki war ceased to be a continuation of politics. But following the nuclear attacks on those cities in August 1945, the USSR entered the war with Japan, this being the continuation of politics aimed at the termination of World War II. There was, after all, no military or strategic necessity to drop those bombs on the Japanese cities; the bombardment was caused by far-reaching political aims on the eve of the Cold War.

M.I. Charles Jr in his *Potsdam Meeting*, wrote that the then US Secretary of State declared that the main advantage of the bomb was not its impact on Japan. Rather, he claimed, the bomb would make the Russians more agreeable in Europe.[2] That was the true political aim of the first nuclear bombings in history.

And if, God forbid, with further proliferation of nuclear weapons they were used again, such a war, regardless of consequences, would be the continuation of a certain extremely irrational murderous policy. Any conscious use of nuclear weapons, whatever the outcome (with the exception of an unauthorized launch) would inevitably be caused by certain political decisions.

Thus, to answer the second of the raised questions, war without politics is not possible.

Therefore, it would be more correct to say that in today's world, war – whether nuclear or conventional – *must not be the means and continuation of politics*.

But if war has broken out, it means that contrary to this imperative, politics have staked all on the military or were forced to resort to it in self-defense. There is a principal difference between the way we wish to see a certain phenomenon or the way it must not be, and its real identity. Nobody objects to the idea of eliminating stealing, and once something is stolen from us, theft in its essence remains theft whatever name we call it.

It is sometimes said that Clausewitz's cliché (that war is a continuation of politics) should be repealed on the grounds that this formula is in itself almost a source of war. But this is tantamount to accusing Newton for the existence of gravitation laws. It is not formulas that produce war, but those objectively existing causes and phenomena that Clausewitz revealed, and these are unavoidable.

To ignore current wars and hostilities and to consider that they have nothing to do with politics and shoot up by themselves, means to lose touch with reality and deprive oneself of the chance to assess events objectively.

It is true that qualitative changes in politics and war assets make the interconnection of policy and war more complicated. On the one hand, they increase the role and responsibility of politics, but on the other hand, nuclear weapons significantly affect politics by drastically altering the very nature of war. Some political analysts rightly emphasize that new armaments are becoming more and more involved in politics and ultimately almost substitute for it. So to prevent war, especially nuclear war, all states need to change their policies, updating them to fit modern reality. Otherwise, Clausewitz's 160-year-old prophecy that sooner or later war would get out of political control and turn into a means of collective suicide, may become a reality. Groundless statements that at present politics and war have nothing in common, mislead people with regard to the main causes of war, deprive them of the chance of influencing ruling circles, and hamper anti-war activities.

2 OLD AND NEW SOURCES OF MILITARY THREATS

In the past, there were always plenty of calls for peace, and denouncements of war. Some of the best minds of humanity fought for peace – take, for instance, Kant's calls for the peaceloving unity of democracies,

Russell's and Einstein's manifestos, to name only a few. 'Eternal peace' has been proclaimed by warring states countless times. Nevertheless, out of four-and-a-half thousand years of history, only about 300 were absolutely peaceful. The rest of the time there have always been armed clashes in various areas of the world.

We have verifiable information about 14,000 wars. Hotbeds of tension exist in many regions of the world in our time as well. And even with the end of the Cold War the world has not become safer – it has become even less safe.

What have been the main causes of all those wars? If we put aside the outer verbal wrapping, the generalized causes of war would be mainly the following: the struggle for territorial gains, natural resources, wealth and manpower (slaves); attempts by conquerors to subdue other nations or even establish world domination; the seizure of colonies, and their expansion, raw materials and spheres of economic and political influence. Nations subjected to aggressive attacks had to fight for their existence, sovereignty and independence.

Deep ideological confrontations have also contributed to war. Religious wars too, have been numerous. During this century, after the 1917 October Revolution in Russia, and during the Cold War, the main global confrontation has been produced by the struggle to impose socialism in all countries, and by the determination of the west 'to crush communism' and establish the total dominance of capitalism all over the world. This is how 'hot' and 'cold' wars have been unleashed up till now.

In generalizing the experiences of the past, some historians have come to the conclusion that no age-long conflict has ever been resolved by military means. And it is true that some armed confrontations produced new ones, leaving the source of the conflict unresolved. But it does not signify that all wars have been senseless, vain and useless in achieving political goals. For many nations, selfless struggle helped retain their independence – Russia was under centuries-long attacks aimed at her division and enslavement. Another example is the formation of the United States in the eighteenth and nineteenth centuries. It is also well known that Germany was formed of small counties unified 'by blood and iron'. It would be no exaggeration to state that humanity would have been thrown decades back, if the anti-Hitler-coalition countries had rejected the use of force with the belief that war was a bad thing, and had not crushed the Fascism of World War II.

In the light of historical experience, it is hard to agree with the statement that 'age-old conflicts are undoubtedly rated as excessive costs of mankind's development – excessive, because they have never been historically unavoidable'.[3] When talking about wars or military conflicts, one presumes that they can be avoided, but we must not forget that in a war one party is the aggressor, the conqueror with the other party on the defence and history still does not provide a clear-cut answer as to what extent war was unavoidable for either side.

W. Thompson, a researcher at the US National Defense University, has worked out some new security principles. He maintains that as the destructive consequences of a conflict using weapons of mass destruction cannot be limited by the frontiers of the sovereign states involved, this type of war cannot be considered as an acceptable sovereign right of a particular state. Arms control has become an international political prerogative. As a consequence of this, the number one principle for security policy in the twenty-first century is no longer based on deterrence, but on guaranteed security for other states; one-sided military buildup should be rejected in favor of mutual safety, with the guaranteed punishment of an aggressor by an international armed force. The security of nations will depend on the extent of their rejection of unilateral military buildup, and on their transition to a policy of cooperation in matters of mutual security in accordance with international treaties. Protection of justice in the world by joint multinational armed forces is the final goal of mutual security, and the gist of national security patterns for the twenty-first century, concludes Thompson.[4]

This approach deserves consideration. Nevertheless, in the past similar slogans and wishes were voiced and so far no state has found it feasible to completely rule out the use of its national military power.

Varying over time, the issue of the application of military power, and especially its limitless application, acquires a completely new dimension which might affect the future of all humankind. That is why the use of armed force should be reconsidered in keeping with new realities. One may assume that with the end of the Cold War global confrontation there is a real chance, for the first time in history, to prevent all wars. 'New political thinking' even proclaims the priority of universal human values above national ones. Emerging contradictions between states should be resolved by political means only. An increasing role for the United Nations (UN) was envisaged with its actual influence on international developments predicted. It was considered that the security of each separate state, and world-wide strategic stability in general, would be ensured not only and not so much by the framework of national security, as through the system of total global security. With this in mind, the Warsaw Treaty was disbanded. Simultaneously, the military security system of former Soviet republics collapsed along with the Soviet Union.

At the same time, actual international events are not developing in this direction.

With the end of the Cold War, the majority of leading states took certain steps to overcome global confrontation, such as the reduction of arms, and maintenance of partnership terms with former socialist countries. Yet as regards international affairs, they are mainly after their own national interests, staking on strengthening their own military security.

For example, in the 1994 'Strategy of the US National Security' report to Congress, there is no word about universal human values. Instead, it clearly states that 'our strategy of national security is based on American values and interests' and even American participation in peacekeeping efforts is considered 'as one of the means to ensure national security, and not as independent operations'.[5]

Under these conditions, the republics which previously made up the Soviet Union have to strengthen their security. Some self-presumptuous political leaders, who lately maintained that they can easily resolve all global contradictions in the world by political means, are now unable to sort out a single problem without violence within a small independent republic. It turns out that the international community is far from being ready for the practical policy of so-called 'new thinking'. What is needed is a sober approach, which would heed existing realities.

Does this mean that future wars will be perpetual concomitants of human history, or will new and more effective ways to prevent them be found?

The answer to this question is that, despite the complexity and controversy of current international developments, there are more possibilities with regard to war prevention than ever before. First, the whole historically accumulated tragic experience of resolving international disputes by military methods, combined with human wisdom, must ultimately erect more reliable blocks to the instigating of wars. Nowadays, no state openly proclaims aggressive goals, and all countries at least make a formal call for peace – and the very fact speaks for itself. The majority of nations are moving towards democratic market economies governed by law. This process must contribute to industrial and informational integration with a reduction in global confrontation, although war between Britain and Argentina over the Falklands, among other wars between capitalist countries, proves that market economies do not help get rid of wars and may even prompt them.

All these facts demonstrate that improved and trustworthy practical steps and mechanisms must be found to prevent wars. And since war is the consequence of policy, which results from political and economic disputes, it is necessary to seek an arsenal of methods for influencing state politics which can forestall any intention to resolve disputes by force. These methods can vary from the activation of relevant public organizations, to strengthening the role of the UN.

Dr Nelson is of the opinion that:

> Security is not only the capability of a state or a group of states, but also an ability to ease threats without, in turn, resorting to threats or use of force. In the next century our security strategy should primarily focus on the latter, which can and must be enhanced by 'taking' security out of the hands of the state ... From today through the coming decade the main goal should be the establishment of regional organizations for collective security.[6]

International legislation defining the military activities of nations needs to be updated. In particular, certain framework agreements (or recommendations) could be produced which would set common requirements for national military doctrines. These doctrines should be aimed at joint security, limiting military activities which would create new military threats for other nations, and declaring the perpetual stability of borders throughout the world. An agreement for Europe, similar to this, was signed in Helsinki. But mutual and global security can be ensured only under guaranteed security of all countries involved, providing them with the required minimum defensive capacity. Therefore, as paradoxical as it may seem, under the present state of international affairs a state's military potential, and the determination by the international community to stop any aggression immediately, have endured as war prevention factors. Nuclear weapons and new types of advanced conventional weaponry will still deter states from unleashing large-scale conflicts. Other factors also now appear which reduce chances to start wars or accomplish aggressive political goals.

In any case, there is a growing understanding of the necessity to demilitarize society. Engels once wrote that 'militarism bears a germ of its own death', and that it would die 'due to the dialectics of its own development'.[7]

But militarism has survived, for it does not exist in a vacuum but rather springs from objective political and economic developments.

Today, we have to take into account that none of the previous sources and causes of war have disappeared; moreover, additional and rather acute social, political, industrial, territorial and ethnic contradictions between states and even regions continue to develop.

George Modelsky and Patrick M. Morgan, political scientists from Washington State University, have done research on the cyclic character of major and world wars, and come to the conclusion, 'We do not consider a new world war inevitable, but we firmly believe that the previous cycles demonstrate that under the ordinary course of political developments the probability of war's outbreak remains, unless a functional substitute for war is found or established.'[8]

In the foreseeable future, the main source of the most acute socio-political and economic contradictions will be the struggle for raw materials, and survival on the planet.

This problem is linked to the more fundamental issue of the ways and means of humankind's further development. The first impression is that the answer to this question has already been given by recent historical experience. The whole world is developing in the capitalist way, and this pattern is being chosen by former socialist countries. Along with that, more and more facts prove that incessant industrial buildup with limitless exploitation of the planet's resources cannot continue infinitely.

Ninety-five per cent of the world's energy is obtained from processing unreplenishable fuels (coal, oil, gas), and only five per cent comes from rivers, tides and solar energy. There are great prospects in the nuclear power industry, but the accident of Chernobyl set limits in this sphere as well. If Russia, China and India alone, leaving out dozens of other countries, started consuming natural resources as generously as the USA, the riches of the planet would quickly become exhausted, and global pollution would be unbearable.

Not only raw materials and water resources, but also forests, arable soils and ecologically clean areas are inexorably shrinking. As a result of anthropogenic erosion alone, more than 25 billion tons of fertile soil are lost annually world-wide. Simultaneously, the world's population is growing. Over the last 20 years it has increased by more than 40 per cent (1.6 billion). At this rate, in 20 years the world population could total seven billion. The price of each square foot of fertile soil, each ounce of potable water or ecologically suitable plot of land, will soar with every passing year making competition for them steeper.

Reputable international scientific organizations and UN experts are unanimous in their opinion that by the turn of the century the ecological factors will become the main cause of political and military conflicts. It is already obvious that the West tries to assign Russia the role of supplier of raw materials, and a dumping site for industrial wastes, but the people of Russia cannot submit to this. Russian scientists and public figures are putting forward various concepts of economic policy, aimed at overcoming Russia's fundamental difficulties and her renewed economic growth.[9]

It is becoming increasingly apparent that the majority of nations will never possess the level of resources which have been under the control of 20 to 25 developed nations; and even if they follow the capitalist way of development they will never reach the industrial level of the advanced nations.

What then shall they do – fight for what their neighbors have? Is that the way to go? There is no clear-cut answer so far. But life makes nations search for an answer in all feasible ways. Many nations, after formally acquiring independence, fell under new forms of industrial and financial dependence from well-off countries. Therefore, their struggle for real independence is ahead, and we can expect another surge of national liberation movements. Struggles for social justice will go on, causing disturbances and civil wars.

Therefore, international communities will sooner or later face problems of the limitation, regulation or qualitative transformation of the scope and nature of industrial output. If the UN, the heads of leading states and the largest transnational corporations and other international organizations do not find ways to coordinate and manage production and consumption, nations will be confronted with the problem of survival (especially in Europe and in the Middle East), so that fighting for resources will escalate, causing corresponding political and economic discord. Here, the prospect of armed confrontation cannot be ruled out. The nature of military threats will largely depend on future political order in the world; on what will replace the bipolar confrontation of superpowers. There are different points of view in this regard.

Some political analysts, statesmen, and public figures deem it inevitable that the future world will be monopolar and under US domination, although this is already unrealizable. Warner has admitted, 'If the US tried to act independently, it would overrun its potential.'[10] This could not happen, not only because of America's steady contraction of her share in world economy, but because even countries linked with America, in particular Germany, Japan and France (let alone China and other powers), would never yield to US domination. If they did, they could not realize their growing economic potential in their national interest. Generally, as history shows, when one state tries to achieve its absolute security, particularly through military buildup, it threatens other nations and heightens competition.

Another group of scientists and political figures maintain that the re-creation of a bipolar world is possible if a number of states in south-east Asia, Eurasia and the Middle East unite around China in order to counter the political and economic expansion of leading western countries headed by the US. This possibility cannot be completely ruled out, but such confrontation – as opposed to cooperation – with industrially developed countries would lead to an increased technological gap, and could instigate a new spiral in the Cold War and even a military showdown.

It appears therefore that under present circumstances the most realistic alternative corresponding to the long-term interests of the international community would be a multipolar world with the US, Western Europe, Russia (along with the CIS countries), China and Japan as centers of military, political and economic power. Objectively, such a process is already under way.

It should be noted that formerly the balance of interests was based on five or six great powers, roughly equally influential. Nowadays, the industrial and military might of leading nations and emerging centers of influence are far from being equal. The US, Western Europe and Japan account for two-thirds of global industrial output.

J. Nye, Director of Harvard International Research Center, maintains that the power setup in international affairs is like a layer-cake. The upper layer – the military one – is mainly monopolar, as no states in the world match the US in military might. The middle layer – the industrial one – has consisted of three segments over the last two decades. And the lowest layer of interdependability is transnational, characterized by diffusion of power.[11] Giving due consideration to the above, the chance cannot be ruled out that before the majority of leading nations finally understand the objective necessity of transferring to a multipolar world, attempts might be made to establish a monopolar or bipolar world, which could lead to a significant complication of the international situation, and displace strategic stability. Besides, any international situation in a multipolar world will be marked by a much greater display of various multinational interests, making it even more complex than the monopolar and bipolar worlds; for the multipolar world might indeed be relatively stable, only if it is based on the balance of interests of individual states. Some political analysts generally consider peace and stability as 'the result of influence of the balance of forces, driven by their own interests'.[12] But the attainment of the balance is feasible only through radical restructuring of the whole system of international relations, and that is highly problematical in the near future.

One of the most important factors affecting international developments and determining the source of contradictions can be the surge in national awareness within numerous nations, and this cannot be reduced to nationalism in its narrow meaning. Out of 170 countries only ten per cent are ethnically homogeneous. Throughout Europe (and not only in Yugoslavia or Russia, but in other relatively politically stable countries like Belgium or Spain), in the Middle East, Asia (Pakistan, India and Afghanistan), national awareness is growing.

The signs are visible. Within the last two years, 22 new countries appeared on the globe. The struggle for sovereignty and independence on a national and ethnic basis continues, and may cause large number of inter-ethnic and internal conflicts. Due to the intensification of antagonisms between advanced and backward countries, the former East–West confrontation may transform into North–South strife.

A group of independent researchers reported to the President of the Russian Federation that they expect 12 new armed conflicts on the territory of Russia and in bordering states. According to their calculations 523,000 people will die in these conflicts, and the number of refugees might reach 21 million or more.[13]

Instability could affect currently stable-looking regions as well. More and more noticeable cracks in multinational states and military alliances appear under pressure of these centrifugal forces.

In this connection, exaggerated hopes of a trouble-free integration process in Europe and NATO (North Atlantic Treaty Organization), in the absence of such a catalyst as 'the Soviet threat', are hardly feasible.

The main threat to international stability may be attempts to revise existing borders. There are hardly any regions without a territorial dispute that has been smoldering for years. There are disputes along the eastern and western borders of Germany, between the Balkan states, the Arab countries and Israel, Iraq and Iran, Afghanistan and Pakistan, China and India, India and Pakistan, China and Japan, Argentina and Great Britain, and so on. The Spratly Islands are claimed by six countries (China, Taiwan, Philippines, Malaysia, Brunei and Vietnam). Similar disputes are abundant. Especially dangerous is the attempt to revise borders established after World War II. Japan's claim over the Kuril Islands, Kohl's statement that he united Germany and that the return of other German lands to Germany is the business of the next chancellor, are very serious. The redrawing of post-World War II borders at least in one or two places may cause an uncontrollable chain reaction of numerous territorial claims among nations, and this would sharpen political and military confrontation.

The Cold War is believed to be over. On the global scale this is true, but not for all corners of the world. A Cold War-style confrontation continues between North and South Korea on the Korean peninsula. The latter hosts American troops. There are reasons to believe that the PDRK (People's Democratic Republic of Korea) is developing nuclear weapons.

The Cold War is not over for Cuba and a few other countries. A lot of 'Cold War roadblocks' persist in international relations, for instance in trade discrimination. All these factors have created and will create further problems among nations.

Any of the above-mentioned 'knots of contradictions' and sources of danger, combined with numerous actual conflicts of national interests in various regions may result in military threats, or in NATO terminology, 'risk factors'. US defense secretary Aspin's report on the new national security policy, submitted in August 1993 to the US president, numbered among other new threats to the US after the collapse of the USSR: 'nuclear proliferation, aggressive ambitions of regional powers [like Iran and Iraq], the peril of failure of democratic reforms in the former USSR; and US economic decline.'[14]

It is very important for the USA, having advanced economic links with almost all countries in the world, to maintain political and economic positions in crucial regions. The deterioration of situations in these regions, or armed conflicts, would affect US interests, and the large number of American troops and bases stationed in various countries may lead to their involvement in armed clashes and local wars. For example, it is not difficult to dispatch troops to Bosnia and Herzegovina, but they may face armed resistance. That would lead to the necessity of their reinforcement and the engagement of more and more new units there, which could increase the conflict as in Vietnam. Moreover, the US could not manage proliferation of conflicts on her own, and thus would be forced to cooperate with other pivotal forces, including Russia.

In the report of the secretary of defense to the president and congress, former secretary Cheney indicated that 'the continuing pull out of the former USSR [from third world countries] means ... that the US gets opportunity to be more discriminate in low-intensity conflicts, enjoying more flexibility and better chances for positive outcome'.[15]

The US military doctrine takes into account not only the peril of local conflicts, but threats of a wider scope as well.

American experts from the National Defense University maintain that major changes in the world with regard to the US would be as follows: future alliances and coalitions would get less reliable; new industrial imperatives would lead to growing instability; relations with allies would get looser as the common threat diminished; former indisputable principles would suddenly become erroneous; new short-term alliances with non-traditional partners would spring up.

The 1994–1998 American military doctrine envisions that the US will strive to 'prevent on the territory of the former Soviet Union the emergence of a new rival which can pose a threat similar to the former USSR' (translated from Russian). S. Pechorov and L. Loginov are right in presuming that the orientation of US policy towards Russia after the signing of the Belovezhskoye Treaty on the creation of the CIS, reminds one of a point in President Wilson's program during the 1918 Brest Peace Treaty epoch, when lamenting over Russia's vastness and homogeneity, he called for reducing her territory to the Middle-Russia Hills, turning the rest of it into a colonial supplier of raw materials for the West.[16]

But the more ambitious the political and strategic goals of a state are, the more military threats it brings upon itself.

There is an impression that some states, the US first among them, are trying to enter the new multipolar world retaining the many attributes of global confrontation. It justifies some experts' reservations that the US does not only possess open (declared) military planning, but has a secret agenda also.[17] Otherwise, it is impossible to explain six scenarios of US troop employment, including the Baltic republics, maintenance of the majority of military bases close to Russia and the CIS republics' borders, numerous infiltrations of US submarines into Russian territorial waters, and other similar actions.

Therefore, as J. Cagley justly observes, 'The US can be a leader practicing high moral principles – ideals of peace, democracy and respect of interests of other nations; or an obstacle on the way to the establishment of a new just world order. It will depend on the priorities chosen by its future leaders.'[18]

As far as the rapidly expanding countries of Germany and Japan are concerned, they are not only expanding their sphere of economic and political influence, but also sending their troops to such remote places as Somalia and Cambodia. Germany and Japan are beginning to make territorial claims upon other countries.

In turn, Japan is concerned by the modernization of Chinese armed forces. Under this pretext Japan is upgrading its armed forces too. If in the past Japanese self-defense forces had the task of protecting the coast and airspace of the country, today's Japanese leaders reckon that Japan should participate in maintaining international peace and order under the UN aegis. Formally, Japan has allocated roughly one per cent of its national budget to defense for many years. However, up till 1973 the Japanese national output grew by ten per cent annually. Her military expenditures grew accordingly, causing concern to Russia and China.

Germany will strive to gradually liberate herself from American control, assuming the role of European security pivot. Margaret Thatcher believes that 'Germany simply is too large and powerful to concede to the status of an ordinary player on European ground'.[19] In Germany, revanchist forces become increasingly active. They are demanding the return of Eastern Prussia, and other eastern regions. Germany may try to reconstitute its eastern lands through economic actions, but potential resistance and clashes of interests among the various population groups in the disputed areas may cause conflicts, and other NATO members could get drawn into it.

Any counteraction to these activities would be considered by Germany or Japan as a threat to their national interest. Besides, these zones of interest are constantly growing. It is no accident that in 1991 Germany insisted on recognition of the independence of Croatia and Slovenia. German leanings towards the Mediterranean are understandable, but this line may happen to change its direction.

For the future, competition among the US, Western Europe and Japan could escalate, although presently the way they have managed to overcome divergence is a good example to follow.

The immense Eurasian continent is turning into a source of perilous contradictions. Russia finds itself in an especially difficult position. Its geopolitical situation has worsened. It has been pressed out of the Baltic and Black Seas, which it had fought to control for centuries.

For Russia, apart from these threats of a general, global nature, other military dangers exist.

The US, and some leading European nations, regardless of their declarations that they no longer consider Russia as their enemy, in fact, remain the main force posing an external threat to Russia. In some near abroad countries nationalist circles proclaim that Russia is their enemy. Relations among the former republics of the Soviet Union remain complicated. Attempts are made to isolate Russia, one example being the plan to create a military-political alliance in the south with Turkey, Pakistan, republics of Central Asia, and the so-called 'Baltic–Black Sea Confederation'.

In the Caucasus, Central Asia (particularly on the Tajik–Afghan border), and in close vicinity to Russia, new conflicts are either ripening or flaring up. War is going on between Armenia and Azerbaijan. Along the whole perimeter of these borders territorial claims are being made, Russian state interests are openly infringed upon, and other states speed up extension of their spheres of influence in the Baltic region, the Balkans and the south. For instance, NATO's 'Bold Game' military exercises in the Baltic zone, and during the 'Dragon Hammer' games on the southern-European theater of military operations in 1992, were aimed at gaining experience in operations on NATO's flanks with the assumption that the 'main risk zone' was Russia and other CIS countries.

Finally, practically all nuclear weapons of the majority of states possessing them, are actually intended to be used against Russia, regardless of declared retargeting. It is true that there are statements in the Russian press nowadays that there is no external threat to Russia whatsoever. But how can we ignore all those threats or at least the nuclear peril? It is still there, as it is for the United States and other western countries. Of particular danger is the threat of nuclear proliferation, and the appearance of other nuclear states. The problems of ensuring nuclear safety in the Ukraine, Belarus and Kazakhstan are of great concern to both NATO and Russia.

The preservation of enormous military might in key global regions is alarming. Despite some downsizing of armed forces in certain countries, the arms race actually continues, especially in the sphere of qualitative upgrading and the development of new types of military equipment. In Europe and many other regions numerous military bases are preserved. NATO member states, and some other countries, still maintain large army, air force and navy groupings. As, for instance, Russian defense minister Pavel Grachev stated, 530 divisions, 42,000 tanks and 12,000 war aircraft have been deployed in close vicinity to the Russian borders.[20]

In the past, third-world countries relied for protection of their interests on one of the superpowers or their allies. Now many of them, having lost such protection, are intensively arming themselves. All these factors represent potential threat to many countries.

Not only these external forces and threats may be sources of confrontations and armed conflicts. Another source of such conflicts are numerous internal problems in various states which are difficult to solve. One of the major reasons determining state policy can be the demographic factor. For instance, UN experts estimate that a stable internal situation in China may be ensured if its population does not exceed 750,000 million. But China's population totals 1,171 million already, and its growth rate is 14 million a year. Such population growth, together with some other bothersome trends, will result in greater tension inside the country. Taking this fact into consideration, Chinese politics and military doctrine are dominated by the thesis which proclaims that as 'the all-round might of the nation' (economy, scientific and technological progress, and military potential) grows, a concept of 'strategic borders and living space' would be implemented, meaning that these frontiers would shift to contain 'the all-round might of the nation' up to the limits inside which the nation could execute its interests in practice, with assistance from its armed forces.[21] And the first frontiers to shift would be those of the weaker neighbouring states. This reasoning thus justifies territorial claims upon other countries, and primarily with regard to south-east Asian states, but it could be aimed at republics of the CIS as well. Internal problems occur also in other states and regions, thus creating potential perils and threats which cannot be overlooked.

Some specialists even think that internal threats are the most dangerous ones. But the sources of domestic conflict often spring up not just from within a country, but can be precipitated from outside.

French General P. Foyar asserts that any notion about defense as the response by military means to an outside aggressor is a return to the past. Defense today must be able to confront internal threats by displaying national unity.[22] Apparently, this is not perceived as applicable to France and other NATO members at all. Nobody has any illusions in this regard. Rather it is a recommendation for other countries. As you can see: some countries receive nothing but threats, while others remain free of such harassment. They are only a threat to themselves.

But there is no need to invent threats. They are born out of a nation's regard for its own sovereignty and national interests, based on deep-rooted vital interests. For example, what are the real objectives of the USA? Obviously not only the 'protection of human rights'. The main objectives of the USA's foreign policy are displayed in the following three directions:

1. Protection of the safe and profitable functioning of American capital in various global regions, which forms the basis of the USA's 'vital interests';

2. Support for political forces in those countries which contribute to promotion of the USA's vital interests, and prevention of political changes which could infringe upon them;

3. Preclusion of the appearance of any state in Eurasia and other regions of the world which could pose a serious challenge to the USA in military and industrial spheres.

The USA makes no secret that these objectives will be pursued using all means, even to the extent of using military force.

But this adamant determination in America's foreign policy will be defied by China's national interests which desires China to be a mighty state with its own expanding zones of influence; and it will encounter Russia, which strives to remain a superpower. Germany, Japan and other prominent nations will protect their place under the sun. These interests, driven by objective necessity, will fight their way through. It is common knowledge that any state has its own national interests which often contradict other nations' interests. But if this or that country meets all the demands of other states without a hitch, what reason is there to attack it? In a case like this there should be no attack, when all dictated conditions are met without effort. On the other hand, when a state starts to protect its national interests, it often encounters counteractions from other nations so that a threat may appear, and an attack cannot be ruled out either. As was said earlier, peoples' growing self-awareness will determine their policy toward national interests.

As far as Russia is concerned, the possibility of its recovery as a great power will not suit many in this world. However, for Russia, the question of whether 'to be or not to be' a great power is not defined by other states which pursue their own interests, and not even by the will of its political leadership. Rather, it is defined by a number of fundamental, objective factors: its historical traditions, its geopolitical situation in Eurasia, and in the world, and, finally, by the reality of the economic, political and spiritual needs which will continue to exist and cannot be abolished.

Without the preservation of Russia's unity, without the maintenance of economic links, building a market economy or democratic state would be impossible, for only capable state power can protect human rights, ensure security of the state, and prevent conflicts. This is consonant with western long-term interests too. But, because it is sometimes ignored, a knot of disputes around Russia and the CIS will be a long-term problem.

A report of the Russian foreign intelligence service dated September 1994, 'Russia – CIS: does the position of the West need any alteration?' reads:

> Some foreign analysts are more and more actively suggesting that irreversibility of the Cold War termination directly depends on the

> preservation of the former Soviet Union in its dissolved shape ... Conclusion is made that the policy of leading western states towards the CIS must be adjusted to keep the status quo, which emerged after the collapse of the Union. These assessments are backed, on the one hand, by grounded fears, that the centripetal process in the Commonwealth might revive a union state in its previous role as the rival of the West, and on the other hand – the clear intention 'to prevent Russia from getting stronger as a world power'.[23]

I think this report contains convincing arguments that the rebirth of a strong Russia, as well as the integration of CIS countries, will not necessarily be accompanied by confrontation with the West. There are ways to avoid this. But it would depend on the extent that Russia, other European countries, and NATO member states, would heed the national interests of other nations and consider the newly emerging realities of the international situation.

In general, competition in the multipolar world among different centers of power, and the aspiration of leading nations to resolve disputes and implement their national interests through limited wars, might lead to an increase of local wars and conflicts. Moreover, unlike the bipolar world, these new conditions would make events harder to control and contain within certain limits. Any state could face a number of unforeseen threats.

One of the root causes of military threat for nations is economic incentives for the arms race, when military-industrial corporations, relevant political and public forces as well as workers, have a stake in it. Military industry, financed by the state budget, has a superprofit considerably above the average. For the majority of arms producers it is five per cent, but for some it is 2,000 per cent.[24]

This system leads to a situation where arms production affects economy and politics. W. Martel and P. Savage in their book *Strategic Nuclear War* write that US policy has always been determined not by a realistic assessment of the international situation, but by the technological level of its nuclear arms, and that technology determines not only weaponry, but the whole way of war thinking.[25] Similar things sometimes happened in the USSR.

Business interests in military production lead to increased arms sales. The ever-growing proliferation of conventional arms beyond any limits, especially in high risk conflict regions such as Afghanistan, Lebanon, Pakistan, Tajikistan, the states of trans-Caucasus, the Middle East, and some African countries – becomes a very dangerous explosive force which could be hard to harness in the future.

Taking it all into consideration, a system of international legislation and standards, which would limit the industrial encouragement and state budgeting of military production, is becoming urgent. Conversion stimulating programs are primarily needed, with increased control over the proliferation of both nuclear and conventional weapons.

In the future a growing influence on international developments will be rendered by transnational corporations (TNC). An assessment in *Monde Diplomatique* states that within the last ten years the annual turnover of 200 major TNCs (mainly American, Japanese, German, French and British) has risen from $3,000 to $5,900 billion, and their share in the global gross national product has reached 27 per cent. Great geopolitical shifts accompany this process. America's share among 200 major TNCs turnover dropped from 43 to 29 per cent throughout 1982 to 1992. At the same time Japan's share grew from 22 to 36 per cent, and Europe's share rose from 28 to 32 per cent. Competition among major state-sponsored TNCs is getting tougher.[26]

While this is taking place, production lines are being increasingly transferred to countries with lower labor costs, thus aggravating the employment situation in capital-exporting countries and unbalancing the global supply and demand.

Victor Kalashnikov, in his comments on F. Clermon and J. Kavanaf's article, justly stressed that the 'rapid expansion of TNCs has become a main factor in forming key global structures within the last 10–15 years. They form a "lining", but are in essence the main content of what is called the international order. The new international order obviously serves as a geopolitical wrapping and indication of the growth and certain reorientation of these developments.'[27] Leading industrialized nations are entering a new stage of the redivision of spheres of influence in Asia and the Pacific region, South America and Africa and in the Middle East. The countries in these regions are preparing to counteract the TNCs in order to protect their own national interests and stimulate industrial and political development.

The most acute outbreak of struggle may be anticipated in Asia and the Pacific. US–Japanese and US–Chinese relations are fraught with processes capable of polarizing Asia as well as other continents. The USA endeavors to expand its influence in Asia and the Pacific region and simultaneously to curb Chinese influence, through bilateral defense treaties with regional states and the deployment of US bases on their territories. This is the reason why not just China but other countries too are disinclined to consider the presence of American troops there as their national security factor.

D. Shambo, a sinologist from London University, emphasizes that Japan could find itself between warring sides, and as its trade conflicts and other problems in relation to the United States aggravate, there is a growing tendency for Japan to side with China.[28] Japan might not support every US effort to bring about political changes in China, for as the experience of the collapse of the USSR demonstrates, it might sharply destabilize the situation in Asia and the Pacific region, severely affecting Japanese economy and the security of the whole region.

S. Huntington, director of Harvard Institute of Strategic Studies, in his article published in *Foreign Affairs* comes to the conclusion that the fundamental sources of conflict in the new world order would be neither ideological nor economic, but would originate out of traditional competition among nations. The main dividing line between peoples and predominant sources of conflict would be of a cultural nature. Nation-states would be the most powerful players in international affairs, but the most significant conflicts on the global scale would arise between nations and groups belonging to different civilizations. Confrontation of civilizations would become a dominant factor in global politics. The future of humanity could be determined through the interaction of seven or eight main civilizations: Western, Confucian, Japanese, Muslim, Slavonic-Orthodox, Latin American and probably African.

In particular, conflicts would escalate along practically the whole northern border of Islamic dominance. Recently, they involved Muslims and Christians in Bosnia, Serbs and Albanians, Bulgarians and Turks in Bulgaria, Occetians and Ingushis, Armenians and Azerbaijanis, Russians and Muslims in the Caucasus and Central Asia, Muslims and Hindus in northern Indian states, Chinese and Turkish-Muslim populations in Sinkiang, and island states in the Pacific.

'The next world war,' asserts Huntington, 'if it breaks out, would be the war between civilizations.'[29]

Future wars and conflicts will I think be not the result of one however weighty factor, but of a complex combination of various sociopolitical, economic, national and religious contradictions. In addition, many aspects of international politics, as described by Huntington, are displayed in the crucial events of recent years.

Each of the analyzed sources of threat may not result in actual peril, since all of them have always existed in international relations in one form or another. But they will be aggravated by a great number of unresolved issues, sharp political, economic and military contradictions in different regions which in sum total, especially when left unchecked, could transform themselves into a serious threat.

So, not only the sources of threats remain, but some new political, economic and religious confrontations may emerge, which cannot necessarily be resolved by peaceful political means. Therefore it cannot be ruled out that in certain circumstances these might result in the application of military force.

Obviously, the end of the Cold War and global confrontation, creates a conducive environment to the resolution of those issues by peaceful means. But the end of the Cold War does not terminate the actual geopolitical positions of different nations or their objective needs and interests. And they cannot remain unresolved indefinitely. Otherwise these interests may find other ways of realization.

Some scientists and public figures choose to disengage themselves from these realities and maintain that opposition to nuclear weapons proves, states and demonstrates to nations and their governments the complete senselessness of military confrontation, and that this would result in a powerful surge of anti-war movements and the total liquidation of arms and armed forces in the world. They think that a revolution in military affairs would transform itself into a bloodless global anti-military revolution. One can only hail these noble wishes and marvelous hopes, but unfortunately they are still idealistic. In reality, objective political, economic and nationalist contradictions continue to exist, and cannot always be overcome by peaceful means. Wars quite often flare up even when both sides try to avoid them. These points cannot be ignored.

3 ON THE FUTURE OF MILITARY–POLITICAL BLOCS

Destruction of the bipolar world and the transition to a multipolar world raises a question of the future of military–political blocs, including NATO.

In the West, as well as in Russia and other CIS countries, NATO is viewed by some people exclusively as a factor of stability, and as a reliable mechanism for securing peace not only in the north Atlantic region, but far outside its borders; extending its zone of responsibility to the territory of eastern European countries, the Baltic, and former USSR states. Others believe that NATO was and still is an instrument for establishing the domination of leading western states (primarily the USA), in Europe.

The third point of view, rejecting the two above-mentioned approaches, asserts that a course maintaining a system of collective security as implemented in OSCE (Organization for Security and Cooperation in Europe) should be followed, because remnants of the past bloc-system only hamper the strengthening of the all-European system of security. The realization of the idea of international security 'from Vancouver to Vladivostok' is considered most advisable.

If we regard world events purely theoretically, and in good faith, then the ideal condition for political restructuring would be the dissolution of all military–political blocs including NATO, or at least their transformation into a politico-economic organization. But we have to emphasize once again that apart from subjective intentions and noble ideas, there are objective realities that dictate somewhat different political imperatives.

First of all, the objective realities of the multipolar world, causing separatism and an aspiration for independence, will at the same time lead to the economic and military–political unification of certain groups of states around emerging centers of economic attraction and power. This in essence is the unity of opposite tendencies, manifesting itself in life.

If we turn to NATO, then not only its opponents but also participants of this alliance see many of its drawbacks and shortcomings. They realize that with the disappearance of the 'threat from the east', the direct necessity for military union ceases to exist.

Membership of NATO requires certain material expenditures, and in one way or another limits national sovereignty. The US secretary of defense insists that European countries be ready to operate outside geographic boundaries of the alliance. This is why active American involvement in military conflicts which could also pull other NATO countries into such conflicts, contrary to their own national interests, has become particularly worrisome. Neither are unified Germany's neighboring states enthusiastic about Germany's excessive strengthening.

Nevertheless, Western European states are holding on to NATO, and other countries persistently seek membership within the organization, considering this should accrue certain benefits for them.

The first benefit is that financial expenditures associated with membership of NATO are more than compensated for by the reliable sustenance of security of each of the states and the union as a whole.

Second, NATO unites Western European countries, and to some extent neutralizes any contradictions emerging among them. It is difficult to say where rivalry between Greece and Turkey over the Cyprus issue and some other matters might have ended, if it had not been for their membership of NATO. This is equally applicable to fears concerning Germany, particularly with French–German competition, because one of the covert reasons for the initiation of NATO was restraining the revival of former German military power. This is why NATO countries are interested in an American presence in Europe.

Third, although presently the USA is the backbone of NATO, it is quite possible that American leadership will be selective. In due time the European Union might acquire greater independence, not necessarily within NATO. The US ambassador to NATO, Robert Hunter, expressed an opinion that theoretically it is possible to imagine that in the distant future Europeans will manage to sustain their security without the USA. But this might only happen after several more decades.

Recently there have been suggestions that Russia and other former

USSR countries might join NATO in order to strengthen partnerships among countries. There is no need for that both from the point of view of these states, and from the point of view of the general military and political situation in the world.

The inexpedience of joining NATO for Russia, was convincingly demonstrated in the document prepared by the Federal Service of External Intelligence. It only needs to be added that in joining NATO Russia will have to subordinate its armed forces to NATO command and control, practically under the command of American generals. Further expansion of NATO is not rational, not only for Russia's interests, but also for the interests of global and European security.

First, inclusion of Russia and other Eurasian countries will render NATO a non-viable organization with new complex problems, because new states will bring in new irresolvable problems.

Second, if Russia, which stretches from Europe to the Pacific ocean, participates in NATO, it becomes an outpost of the West in the east, and will have to oppose China and other countries situated in south-east Asia. If we hypothetically assume these states also might join NATO, then the question is, how far could NATO extend? Limitless expansion of NATO would lead to a situation where in due time it would claim to substitute for the UN. If at present at least the semblance of equality of countries and peoples is maintained, then given NATO leadership, all countries will be subordinated to a certain group of leading states. This will inevitably lead to counteraction from other countries, including western ones, creating new beds of confrontation.

Even the partial extension of NATO due to the inclusion of Eastern European and Baltic states, will result in the advancement of NATO's boundaries to the east, and will lead to the further isolation of Russia. This might cause a discontinuation of the Paris Agreement on the reduction of conventional weapons, as well as disrupt the correlation of forces and of any remaining weapons and equipment stipulated by this agreement. One should not forget that the USSR, on giving consent to the unification of Germany and the withdrawal of its forces from central Europe, was promised by NATO countries that this move will not be used for shifting the strategic-military balance in the region to the benefit of these states. But these promises, too, are being broken. This is why the former prime minister of the FRG (Federal Republic of Germany) has said: 'If I were a Russian marshal, I would have considered the expansion of NATO first from Elbe to Oder and then to the eastern borders of Poland as a provocation and threat to Russia.' One circumstance of note is the fact that NATO leaders signed a document 'Partnership for Peace' in January 1994 in Brussels, without even thinking it necessary to discuss its conditions or even consult the countries invited to participate in 'partnership'. Moves like this increase NATO's possibilities of domination even more. If Russia joins NATO, it will be limited in external political and military respects and will lose freedom of action, at least for such independent actions as in the settlement of conflict in Sarajevo.

With regards to the above-mentioned circumstances, it should be said that unlimited extension of the military–political structures of NATO will lead to a destabilization in the international situation, and hamper the strengthening of international cooperation.

If the extension of NATO and Russian involvement in it does not make sense, then the program 'Partnership for Peace' is also unnecessary, because the main idea behind it is a phased extension of NATO.

This is explained by several circumstances.

First, the 'Partnership for Peace' program launched by the USA when considered from the point of view of narrow NATO interests, is a far-seeing step. It puts Russia into a difficult position. By refusing to participate in the partnership program Russia seemingly juxtaposes itself to other states hurrying to accept the program, and might find itself in a certain isolation.

If Russia joins the program within the set frames, it will have to limit its sovereignty in military–political matters, because regardless of the fact that the partnership program formally stipulates the conclusion of an agreement with each state individually, it sets rather strict general requirements for military planning, budget, the standardization of armaments and joint exercises.

Besides, when the day comes to shift from a partner to a member of NATO, Russia will face this dilemma – should it apply for membership, or experience the threat of isolation?

Second, a priority orientation for Russia and other former USSR republics toward joining the NATO partnership program disrupts already-weak system of collective defense and military cooperation within the CIS.

By following all the conditions of the 'Partnership for Peace' program in the areas of military planning development of equipment and training of the armed forces will increase NATO's influence in these areas and will gradually diminish, and then reduce to zero Russia's role in coordination of military cooperation within the CIS, and will compel these countries to the preferential adoption of NATO weapon systems, leading to the even greater destruction of the defense industry in Russia, pushing it out of world arms markets.

Actually, with all adherence to principles of the equality of states, Russia, possessing huge nuclear arsenals and having a number of international obligations of global importance, cannot accept the 'Partnership for Peace' program on the same conditions as say Estonia or Latvia, because these countries are more interested in guarantees of security from NATO after they become members, than in cooperation. The threat they are seeking safety from is meant to come from Russia.

In view of this direction, adoption of the partnership program for Eastern European and Baltic states is just a delay in joining NATO. Discussing the document on the partnership program adopted in Brussels, the US Senate in its special resolution called for prompt admission of Eastern European states into NATO.

As far as Russia is concerned, it does not need anybody's guarantees of security, and should ensure its defense independently. For Eastern European and Baltic states it would be expedient to procure joint guarantees both from NATO (or the USA) and Russia.

Participation in the 'Partnership for Peace' program is rather burdensome in financial respects. For example, it will cost Poland 400 million US dollars. It was reported in the newspapers that the Lithuanian government, having joined the partnership program, decided to start by allocating one company in conformity with NATO's requirements, assigned for peacekeeping operations. It turned out to cost ten million US dollars, a quarter of the annual military budget of the country.[30]

At the same time it is evident that despite all of the listed negative aspects of the 'Partnership for Peace' program, Russia cannot disregard it and stay outside the integration processes in Europe, including matters of military cooperation. Within the frames of national interest, Russia must try as much as possible to neutralize negative points of the program, and to use it to full benefit.

In this regard it is thought to be advisable for Russia:

1. To support in principle the idea of partnership and military cooperation in Europe, establish necessary contacts, and to undertake joint actions on securing safety with the military–political structures of NATO.

At the same time it is essential to take into account that the military organization of NATO, designed to repulse the so-called 'Soviet threat', has practically outlived itself and sooner or later will have to be transformed with regard to new conditions in the international situation. Only the North Atlantic Cooperation Council might be preserved, if it becomes an independent organization closely connected with the OSCE; this makes the extension of NATO and its zones of responsibility even more unjustifiable.

In the long term, the main emphasis in Europe should be on strengthening the corresponding structure of the OSCE, transforming it into an independent all-European military–political organization.

Within the multipolar world, basing itself on the balance of interests of different regions and states, Russia can actively participate in integration processes, and establish military cooperation with the corresponding international organizations of Europe, America, Asia, the Middle East – and first of all within the CIS.

2. At this stage in the development of the international situation it is logical to conclude corresponding agreements with NATO on partnership and military cooperation, but not within a formal, strictly defined 'Partnership for Peace' program, but through agreements on broad, equal interaction on concrete issues of partnership and military cooperation, which are mutually beneficial.

The practice envisaged by the 'Partnership for Peace' program, where NATO requires other partners to disclose military plans without disclosing their own, and to adjust military theory and practice to NATO standards without borrowing from the useful practices of other countries, will only impoverish the development of military affairs and make it one-sided.

Such a partnership ought to provide mutual enrichment of experience and knowledge in the area of military planning, the exchange of novel technologies, the development of arms and organizational structures, and the combat and tactical training of troops.

3. In concluding agreements on military cooperation, Russia should reject unacceptable conditions for military planning and budgeting, and especially the creation of peacekeeping forces under the aegis of NATO. Peacekeeping functions could be vested in the OSCE or the UN, which in their turn might assign certain peacekeeping missions to certain states or groups of states.

Thus, partnership and cooperation are necessary, though not on a basis of pre-set standard conditions, but on a broad, mutually-beneficial and equal basis, with due regard to the national security interests of all the participating states, and a general, mutually-dependent security in Europe and other regions of the world.

Taking everything into account, Russia and other CIS countries should establish political and military cooperation with NATO on a bilateral and multilateral basis within the OSCE.

In forming the multipolar world, the possibility of creating other regional military–political and economic alliances in the CIS, Middle East, southeast Asia and other regions is not excluded.

At the conference of Russian and Japanese security experts in Moscow in October 1993, the director of the Japanese Institute of Defense Research, Hironaka, noting the necessity for such organizations as NATO, the EU (European Union) and the OSCE, has spoken in favor of creating similar organizations in the east. The regional ASEAN (Association of South-East Asian Nations) forum comprising 18 countries, was also mentioned.[31]

Some Asian countries view with apprehension the possibility of the participation of such superpowers as Russia, China, the USA and Japan, being afraid of their excessive influence. To my mind, the involvement of superpowers might have a positive consequence by stimulating cooperation among them. As President B. Yeltsin noted at the UN General Assembly 'The world system of the future will really become stable only if we manage to find effective mechanisms of solving problems at each level: national, regional and global.'[32]

But the main thing is to persistently and successively strengthen trust among peoples and take into account the balance of interests of different states and regions. If domination by two superpowers is to be substituted by domination under any seemingly noble pretexts of one superpower or a group of leading states, and if countries that broke out of the Cold War are to be divided into winners and losers (the attitude to losers is well known), then nothing will change in the world, and new sources of conflict will emerge, giving rise to wars and confrontations.

But we do, in fact, have the possibility of avoiding a renewal of global confrontation. To turn this possibility into reality, countries need not only to declare but to change their course of politics. Efforts need to be directed at the creation of an all-round global security, without infringing the interests and safety of any country.

Notwithstanding the imperfections of the world, one should hope that previous tragic experiences have not been in vain, and that national governments, peoples and international communities in general (including the UN), will somehow work out mechanisms for settling confrontations without war.

As recent experience shows, one should not underestimate the dangers of military conflicts, or excessively easy interference in the internal conflicts of some countries. Local wars and military conflicts, irrespective of the will of their initiators, can grow and take on an absolutely different scale. That is why it is important for nations to take into account the interests of other countries, and not try to bring inter-state affairs to an extreme tension.

The interests of the world community require an increased role for the United Nations in the prevention of wars and military conflicts. The time has come to rethink a lot of things about the activity of this most important organization. Recently, a contradiction has been in evidence when on the one hand the UN charter forbids the use or threat of force in international affairs and demands they be settled peacefully, and when on the other hand military force is increasingly being used on behalf of the UN, creating conditions for abuse. And the use of force is sometimes resorted to before all political possibilities of settlement are exhausted. The absence of any prior political settlement puts peacekeeping forces in a difficult position, because as Talleyrand said 'a lot can be done with bayonets, but it is not possible to sit on them'.

In this connection, some specifications to the UN charter seem to be essential.

Recently, an excessively broad interpretation of peacemaking actions has been evident. At the Moscow international conference, held at the Center of Political and International Studies in November 1993 with representatives from NATO, some speakers referred to peacekeeping operations as not only the assignment of units under the UN flag to Yugoslavia, Kampuchea and Somalia; but as including the war in Korea in the 1950s, the Gulf War in 1991, and some other operations. Under this approach, one can easily view World War II as a peacemaking action. The danger of this broad interpretation is, that under the influence of a number of leading states, the use of force on behalf of the international community against separate states can be legitimized. Still, war is war, and peacemaking operations are an absolutely different thing. The latter should be connected not to the large-scale application of military force, but to actions directed towards promoting settlements of military conflicts.

In general, it should be said that the UN, as well as any other international organization, can function effectively only if the member-states are deeply interested in cooperation among them.

But so far the majority of nations, waiting for better things, deem it necessary to take care of their own defense as far as they can.

The reality is, that only states strong in economic and military terms are respected.

Why is the USA taken into account all over the world, why are its interests extending so far, when no other state can afford to do likewise? This is not explained by any specific qualities or intentions of these countries' leaders, but by only one circumstance – the USA is the most powerful state in economic and military respects. And all nations need to take this fact into account, and draw appropriate conclusions from it.

NOTES

1. G.A. Arbatov, *Svidetel'stvo sovremennika* (Moscow: Mezhdunarodnye otnosheniya, 1991) p. 345.
2. *Za Rubezhom* (1985) No. 30, p. 17.
3. E.B. Chernyak, *Vekovye konflikty* (Moscow: Mezhdunarodnye otnosheniya, 1988).
4. Wayne C. Thompson, 'Redefining Strategy 2000 and Beyond', in *Essays on Strategy* (Washington, DC: National Defense University Press, 1992) pp. 105–24.
5. *Krasnaya Zvezda* (23 Sept. 1994).
6. Dr Nelson in *Mezhdunarodnaya zhizn'*, No. 9 (1993) pp. 58–9.
7. K. Marx and F. Engels, *Sochineniya, izdanie 2-e*, Vol. 20, p. 175.
8. G. Modelsky and P.M. Morgan, 'Understanding Global War', *Journal of Conflict Resolution*, Vol. 25 (1985) p. 415.
9. For example see: S.Y. Glaz'ev, *Teoriya dolgosrochnogo tekhniko-ekonomicheskogo razvitiya* (Moscow: Vladar, 1993).

10. Warner in *Compass* (1992) No. 75.
11. J. Nye in *SShA: Ekonomika, politika, ideologiya*, No. 11 (1992) p. 6.
12. *Mezhdunarodnaya zhizn'*, No. 9 (1993) p. 52.
13. *Pravda* (9 February 1994).
14. *Krasnaya Zvezda* (2 Sept. 1993) (translated from Russian).
15. Report of the Secretary of Defense to the President and Congress, p. 12.
16. *Voyennyy vestnik*, No. 18 (1992) p.3.
17. *SShA: Ekonomika, politika, ideologiya*, No. 11 (1992) p. 47.
18. J. Cagley in *Ethnics and International Affairs*, No. 46 (1992) pp. 36–7.
19. *Za Rubezhom*, No. 45, 1993.
20. *Voyennaya mysl'*, No. 6 (1993), p. 4.
21. *Voyennaya mysl'*, No. 1 (1993), p. 67.
22. *Révue de défense nationale*, No. 4 (1991) p. 20.
23. *Nezavisimaya gazeta* (22 Sept. 1994).
24. R.A. Farmazyan, *Voennaya ekonomicheskaya politika amerikanskogo imperializma*, (Moscow: Mysl', 1983) p. 152.
25. *Novye knigi za rubezhom*, No. 8 (1987) p. 55.
26. *Monde diplomatique*, No. 3 (1994).
27. *Nezavisimaya gazeta* (1 April 1994).
28. *Za Rubezhom*, No. 12 (1725) 1994.
29. Cited from *Svobodnaya mysl'*, No. 17–18 (1993) p. 22.
30. *Krasnaya zvezda* (29 March 1994).
31. *Krasnaya zvezda* (15 Oct. 1993).
32. *Rossiyskie vesti* (28 Sept. 1994).

CHAPTER 2

MILITARY–TECHNICAL PROGRESS AND ITS INFLUENCE ON THE NATURE OF ARMED STRUGGLE

> Few things depend on economic conditions so much as an army or fleet … The whole organization and method of combat used by them, as well as their victories and defeats, appear to be dependent on material, i.e., economic, conditions – on human material and weaponry.
>
> F. Engels

Throughout time one of the major factors determining the nature of war and war-fighting methods, was the invention of new weapons and the modification and perfection of existing arms and military equipment. For many decades, former socialist countries proceeded from Engels' assumption that perfection of weaponry and equipment determines development of strategy and tactics. This assumption was not denied in other countries too. Basically, notions about the development of military art in accordance with changes in weaponry will retain their importance even into the future. But now, when military–technical progress is accelerating, when the cost of arms is ever-increasing and putting even more of a heavy burden on the economy, this assumption is already insufficient. Under modern conditions and even more so for the future, military science should itself define the general directions of arms development, to develop more specific operational-strategic and tactical requirements (a description of the weaponry and equipment of the future). In all circumstances the influence of strategic assumptions will be constantly growing.

In this connection, the main point to be taken into account by all armies of the world is the all-round increase of efficiency in all measures, relating to the construction of the armed forces, to the stricter, more consistent, and thorough account of military-economic and strategic assumptions. Proceeding from the above stated facts, even the most underdeveloped countries will not enter the outlined path of the arms race, but will aspire to select a way appropriate to their economic abilities and strategic objectives, and to assert favorable conditions of struggle to their military and technical advantage.

It will be especially important to prevent dispersion of forces and resources, under the aim of developing all kinds of weapons and equipment, which could then be used in other armies. In necessary cases, it will be expedient to undertake a calculated risk and to pass up some forms of weaponry, in order to achieve a really significant concentration of technological and industrial efforts on the development of weapons of key importance, which will neutralize or break the long-term programs of other countries in the attainment of military superiority. Even with military doctrines of a defensive character, it is most expedient to develop systems of attack, and to have an increased number of multipurpose systems.

Military–technical policy within the USA envisages the preferential financing of RDTE (research, development, testing and engineering), aiming at the buildup of US technological leadership in the field of the qualitative perfection of arms; increased attention to prototypes with demonstration of weapons and military equipment; transition from the prototype stage to a large series production only after a complete and extended cycle of tests, and even then not in all cases; transition from the production of brand new types of arms to the modernization and improvements of old types of equipment.[1]

In general, a more active military–technical policy, capable of dictating the conditions of the struggle for military and technical superiority in some major directions, will win.

The following factors will produce the greatest long-term effects on the development of arms and equipment:

- the absence of global confrontation by the two military and political blocs, the reduction of the possibility of world war, and the resulting new nature of armed struggle;
- the realistic estimation of possible military threats, armed conflicts, wars and diverse tasks, which armed forces will need to fulfill;
- concluded and prospective new agreements on the reduction and restriction of nuclear and conventional weapons;
- the geostrategic position of nations, and the degree of their involvement in various military blocs;
- economic difficulties, and financial restrictions over military expenditure, which are more and more difficult to justify now that the Cold War is over;
- the increased importance of the efficiency criterion combined with an increasing need for dual-purpose equipment;
- the development of new technology, in order to greatly improve the qualitative parameters and efficiency of new modifications of weapons.

The main emphasis will be on the creation of broad scientific and technological arenas to develop promising models as refined designs and technologies of the major subsystems; and to develop model studies of different versions of new weapons systems for construction. The influence of military science on the development of weapons and military equipment is increasing.

Under the influence of these factors, the basic directions and tendencies of arms and equipment development could be as follows.

1. Information processes will acquire decisive importance. Proceeding from this assumption, priority will be given to the systems and weapons complexes in which increased efficiency and combat capabilities of weapons are attained. Reconnaissance, communications, radio-electronic warfare, and automated control systems of troops and weapons will play a special role.
2. Major attention will be paid to the perfection of conventional precision weapons, with primary development not of defensive, but of offensive assault means, as being the most effective and economic.
3. In the development of strategic means, the three-fold weight of mobile components, primarily missile submarines and strategic aviation, plus long-range cruise missiles, will steadily increase.
4. More attention will be paid to increasing the survival of weapons systems and personnel through the utilization of improved camouflage, through fire and radioelectronic destruction of the enemy, and increased maneuverability and mobility of combat equipment and troops.
5. The high cost of operations, weapons, and complex equipment requires a more intensive development of facilities for the combat training of troops; especially those simulating shooting, missile launching, the flying of aircraft, the navigation of ships, and other kinds of equipment.

The main tendency in arms development, which will become more apparent in the long term, is not the creation of individual types of weapons, but the design of weapons systems. Under modern conditions, the capabilities of even the most perfect weapons, the best-equipped groupings of troops, air and naval forces, are most fully implemented when integrated into a unified combat system with highly organized functional properties of intelligence, organization, observation, control, secrecy, and accuracy of operation. The main efforts in the struggle with the enemy will be directed not towards the physical destruction of each unit of weaponry, but towards the destruction of their common information space, sources of intelligence, channels of navigation, and control systems of communication and targeting in general. Accordingly, planning systems of weapons development and the hardware of armies and naval forces will be improved.

1 TECHNOLOGICAL PROGRESS IN ARMS DEVELOPMENT

It is safe to say that readiness to use the achievements of science and technology in the military arts will remain. But the demands of military art will certainly continue to stimulate developments in many branches of science and technology, especially those of nuclear physics, chemistry, exploration of space, in ballistics, theory of probability, radar, automated systems, nuclear propulsion for naval forces, aircraft construction, aerodynamics, missile construction and so on.

If prior to World War II applied sciences bore the greatest influence on arms development, so now and in the long run influences of discoveries in basic sciences will be on the increase. Time spans between discoveries and their technical realizations are reducing sharply; while the role not only of physics, chemistry and mathematics, but also of the biological sciences, in particular bionics, is increasing.

More attention is being given to sciences connected with the study of personality, with the aim, on the one hand, to develop a means to promote the enhancement of inner moral-psychological stability to the stress of armed struggles of the future; and on the other hand to develop means and methods to influence the central nervous system and mentality of people, with the aim of amplifying fear and paralyzing the will.

Overall, from a military–technical point of view, we can expect the transition of the armed forces of industrially advanced countries on to the new technological basis. A qualitative 'technological explosion', with widespread computerization and creation of artificial intelligence, further developments in microelectronics, 'smart' intelligent weapons, and the introduction of automated systems and robotics into the processes of management, can all significantly change the material basis of armed struggle. The opportunity to create a completely new weapon ('superweapon'), based on new physical principles cannot be excluded. In the world of elementary particles there are still opportunities to discover new, even more powerful sources of energy. The production of atomic energy is constantly improving, especially its application for the navy. Work is going on in the field of the application of light generators and amplifiers, using interaction processes of elementary particles for the concentration of beam energy in communication and location; the development of laser and similar weapons based on the utilization of various kinds of electromagnetic radiation. Jet engines of the new type are designed to power superhighspeed aircraft and missiles of all classes.

The most important factor in the development of all kinds of arms is in increasing their maximum range – the range and accuracy of missiles, and the speed and flight ceiling of combat and transport aircraft. Strategic weapons (missiles and bombers) can now reach practically any point on the globe. The maximum range of air and rocket-artillery systems of operational-tactical assignments are also continuously being increased. Cruise missiles, which can be launched from land, air or water have grown into completely new weapons. The difficulty of their detection, small size, high accuracy and maneuverability make them extremely effective weapons.

The time factor has always played a key role in war. But now its importance is increasing with unprecedented speed, and time specifications are compressed to the limit. Increasing rates of fire of automatic weapons, the flight speed of shells and rockets, the movement of combat weapons carriers (aircraft, tanks, APCs [armoured personnel carriers], self-propelled guns, motor vehicles, air defense complexes, surface ships and submarines), all provide the speed of application of these combat means and at the same time accelerate all the processes of the preparation and conducting of combat, and so speed up the development of combat actions and operations. And the combat application of strategic nuclear forces (for example, it takes 25–30 minutes for a ballistic missile to cover a distance of 10–12,000 km), space antimissile and air defense means – the detection and warning of an enemy missile launch, the issuance and execution of orders on application of your own means – are already estimated in minutes and even by fractions of seconds. All this makes armed struggle even more dynamic and transient.

Speed and range of weapons application requires appropriate speeds of communications and administrative responses, an accelerated coordination of efforts, and the interaction of the forces and means involved in armed struggle.

The increased power of shells, bombs, rockets and other kinds of ammunition allow a shorter timespan and greater accuracy for hitting various objects and whole groups of troops, aircraft and naval forces of the enemy. Conventional precision ammunitions come in their effectiveness of application closer to low-yield nuclear devices. As far as the destructive effects of nuclear weapons are concerned – they are well known. Suffice it to say that the capacity of nuclear ammunition of 10 megatons is equivalent to the total capacity of all ammunition used in former wars when gunpowder was used, not to mention penetrating radiation, light radiation, and the radioactive contamination of the terrain, that accompanies a nuclear explosion.

Inventing ways and means for the neutralization of nuclear weapons platforms on ground, sea and air and in space, can become a turning point in the struggle of attack and defense. Scientific discoveries in chemistry have created synthetic materials, rare metals and alloys, which are widely used in the production of various types of equipment; including radio-electronics, missiles, aircraft, armored vehicles, artillery and reconnaissance devices.

Developments in cybernetics, computers and communication facilities of the new generation transform not only their material basis, but also the content and methods of command and control; allowing collection and systematization of situation data in shorter time; to estimate and forecast its development in several variants; in mere seconds to set up tasks for troops and activate weapons.

Various kinds of so-called 'non-lethal weapons', based on the use of electromagnetic pulse generators, are being intensively developed. Without killing people, such weapons can disrupt the functioning of telephones, radar stations, computers and other means of communication, navigation, aim and control systems.

There are also 'combustion inhibitors' halting car engines; also chemicals that destroy the tyres of motor vehicles and aircraft, if spread on roads and airfields. Stress is placed on the fact that utilization of means like these will paralyze the personnel and equipment of the opponent, without destroying them.

The absence of international legal restrictions makes it possible to create completely new kinds of mass destruction weapons. Especially dangerous could be methods designed to exert an influence on human psychology, with delayed results, which might not be manifested immediately after employment, but only months or years later. The achievements of genetic engineering and molecular biology of microorganisms and higher organisms, create the preconditions for production of human toxin, which can be used as a genetic weapon capable of disrupting the normal genetic hereditary functioning of living organisms.

The American medical specialist, R. Hammershlag, believes that the study of natural and genetic distinctions between people, and their subtle biochemical structures, has revealed the possibility of creating so-called ethnic weapons, the selective application of which (by differentiating blood groups, skin pigmentation and genomes) will result in losses of up to 25–30 per cent of certain groups of the population.

High frequency electromagnetic waves and infrasound waves can produce a powerful hazardous effects on humans. Low frequency fluctuations can cause epilepsy and other functional disorders, including numerous cases of myocardial infarction among military personnel and the civilian population.

The invention of 'geophysical weapons' capable of causing various earthquakes over certain regions, tsunami, destruction of the ozone layer, or downpours, is considered quite feasible. Similar examples occurred during the Vietnam War – irrigation facilities were destroyed, causing floods.

Methods and means of information war, aimed at influencing the consciousness and feelings of the population and at subversion of the state from within, are becoming more and more sophisticated. The systematic broadcasting of psychologically- and ideologically-biased materials of a provocative nature, mixing partially truthful and false items of information; the piling up of artificially exaggerated sensational facts that have impact on human imagination, can all result in a mass psychosis, despair and feelings of doom and undermine trust in the government and armed forces; and, in general, lead to the destabilization of the situation in those countries, which become objects of information warfare, creating a fruitful soil for actions of the enemy.

The practice of the extremely prejudiced and free manipulation by the mass media of public consciousness has become the subject of two sad anecdotes ... Napoleon came back to life and was looking through the newspapers during a military parade in Moscow. When asked by journalists: 'If you'd had missiles and tanks like the ones displayed at this parade, would you have lost the battle of Waterloo?' Napoleon answered: 'I am thinking of an absolutely different thing. If I had had newspapers like yours, nobody would ever have known that I lost that battle.'

In 1898 one of the magnates of the US press, William Randolph Hearst, summoned his employees and gave them an unexpected order:

'Publish the photos of the beginning of the war between the USA and Spain.'

'But the war has not yet begun,' answered the startled journalists.

'Publish the photos, and it will begin,' Hearst answered.

This story is considered an anecdote, but nevertheless, falsified provocative photos helped to undermine the relations between these two countries. Even today, many conflicts in former Yugoslavia, in the Caucasian republics, central Asia and other regions, are to a considerable degree inflamed by the mass media, which are controlled by certain political and Mafia groups, and the special forces in a number of these states.

General V. Belous, speaking about new methods of mass destruction and recalling the above-mentioned passage on Hearst, noticed that the utilization of information and other new means for the achievement of political, economic and ideological goals, creates an opportunity for the transition from direct armed struggle to ways of latent, undeclared war.[2]

Greater importance for ongoing armed struggles in local military conflicts is attributed to the new generation of non-lethal weapons which affect the resistance of armed formations, and do not cause great damage to the civilian population. These include: lasers, microwaves, light and electromagnetic pulses, microorganisms, chemical substances, computer viruses and other means. They can be used for various actions, starting with the dispersal of demonstrations, and finishing up with the isolation of whole regions affected by mass disorders and armed confrontations.

For instance, in the USA and other countries 'exotic technologies' and weapons systems are being developed. As was reported in the American press[3] these might be powerful lasers capable of disorienting pilots and disrupting navigation and control devices. The lasers can be transformed into generators of less powerful pulses, which do not kill, but only knock out live targets; microwaves and non-nuclear electromagnetic pulses, disrupting the operation of radiostations, computers and electronic systems; special liquids, delivered to the target as aerosols, which in only minutes can make metal fragile. The aerosols can be used by peacekeeping forces too, in order to establish a barrier between opposing sides.

Sound wave generators capable of producing noise that causes physical pain are also being developed; also red and dark blue light pulses that cause nausea; and a means of spreading filthy smells to disable the actions of troops. Powerful flare lamps which can lead to temporary blindness of rioters are being designed.

Also envisaged are the spread of microorganisms able to eat away electricity cables; particles of a self-igniting substance that can set engines on fire if it gets in through air intakes; and aerosol 'slippers' and 'stickers' that can make roads impassable. A special chemical composition can condense diesel fuel and petrol, rendering it unsuitable for use.

The invention of a police rifle which can fire so-called 'bean bags' and knock a person down (they are considered safer than rubber bullets) is reported. The utilization of unusual kinds of foam with great sticking force is also possible. Terrorist or bandit formations can be targeted with this composition, which turns into a sticky mass when exposed to air. Another type of foam forms an avalanche of very dense soap bubbles, which deprives the opponent of hearing, speech and movement.

The existence of metallic nets for combat and transport vehicles, and thin ones for personnel is also known. They can be fired in containers the size of a large bottle, exploding over the target. After destruction of the containers the nets spread and cover the target. More data is reported about other similar means.

Space and Nuclear Arms

The nature of armed struggle will be heavily influenced by space systems.

An agreement on the principles of activity of nations in the exploration and use of cosmic space was signed by the USSR, USA and Great Britain as far back as 1967. But developments in military space systems still go on, including some programs that infringe established principles (sometimes under the disguise of other programs). It is customary to divide space assets into three groups:

- space systems for the support of combat actions (reconnaissance systems, early warning and detection systems of enemy missile launch, communications systems, TV, navigation, topogeophysical and meteorological support);
- antimissile systems: from space interceptors to neutron and X-ray lasers, as well as systems designed to strike from outer space against targets on Earth;
- systems of defense and attack in space, including the means of detection, recognition and destruction of satellites in space.

The last two directions have found their most complete implementation in the US program of so-called 'strategic defense initiative' (SDI).

The unfeasibility of the US SDI program was obvious from the very beginning of its announcement. Just as clear, was its main purpose. S. Belikov, R. Sagdeev and A. Kokoshin wrote about it. 'In the directives in the field of defense for the years 1984–1988' the US Department of Defense specified that the United States should prepare 'to effectively conduct war' from outer space. And further on, 'it is necessary to direct military confrontation with the USSR into new areas and thus to make all previous Soviet defense expenditures senseless and render the whole Soviet weaponry outdated.'[4] This aim was not completely achieved. The USSR did not enter this race, and economically it was already incapable of undertaking programs similar to SDI.

However, as the reduction of strategic nuclear forces proceeds, the reality of the creation of an antimissile defense system able to repel limited nuclear enemy attack increases. The side which possesses such a defense and space system acquires huge advantage and the opportunity for military superiority, while the other side decreases its opportunity for effective nuclear retaliation.

Furthermore, space systems which were supposed to be created within the framework of the SDI program were disguised as defensive, but essentially were exclusively offensive. They were to be used for the destruction of enemy early-warning and communication satellites, to ensure a surprise nuclear attack and to deprive the opponent of the chance of timely retaliation. The plan was to design space strike systems with an operational range of up to 4–5,000 km, able to destroy major strategic targets on the ground in several seconds – which is especially important in the delivery of a preemptive strike. Space systems will play a large role in the support of combat actions of all kinds of armed forces.

Troops on a battlefield will receive real-time situation reports not from higher HQs through the chain of command, but directly from satellites or a central command post. In the long term, it is possible that commanders of advanced units and even tank crews and APCs will be able to monitor the formations of enemy and friendly units, on personal mini-displays. The necessary information will be sent directly into pilot cockpits or even to missile warheads, launched from long distances, as they approach their targets.

Further perfection of nuclear weapons is aimed at increased accuracy against military objects, the ability of nuclear missiles to overcome systems of antimissile defense, the creation of third-generation nuclear ammunitions with reduced radioactive emission, super-EMP (electro-magnetic pulse), which employs an electromagnetic pulse of high voltage capable of putting out of action all electronic systems of the enemy, and which affects ground-based missile complexes, radar-tracking stations, communication facilities, methods of automation, and counter-electronic warfare.

Under the reduction of arms in accordance with the agreements in START-2 (strategic arms reduction talks), by the end of this century the weight and importance of air and naval components in the system of strategic nuclear forces will increase. It is possible to forecast that Russia, China and other states will not be capable, especially in the qualitative aspect, to create strategic air and naval forces equivalent to those of the USA. Therefore, in general, the domination of US strategic nuclear forces will steadily grow, despite their apparent quantitative reduction. The creation of space systems with nuclear and laser weapons and various types of electromagnetic radiation, is dangerous not only because control of these systems is possible only through automated control systems, which means that hazardous powers of mass destruction are practically completely out of human control, but also because even the slightest technical failures and errors (which are not uncommon) can result in unauthorized employment of nuclear weapons. All this creates a completely new and dangerous military and strategic situation in the world.

This is why it makes sense to listen to advice given by such authoritative experts on military and strategic matters as Robert McNamara, who said in 1986 that the attempts to attain nuclear advantage are not only useless, but are also dangerous. 'Stable balance at a possibly lower level should become the purpose. Our technological "spine" should be used at full capacity, to strengthen our safety, but not so as to threaten the stability of deterrence.' We should not forget the warning given by Winston Churchill that 'the stone-age can return on the shining wings of science'.[5] Latest reports testify that the USA has given up the realization of a full-scale SDI program and even of the outer-space striking component, re-orienting SDI into ABM (antiballistic missiles) on USA territory and the theater of war. But so far, each new US administration has had its own approach. And this can change.

Therefore, in the interests of the whole international community, it is not enough to rely only on the common sense of individual rulers, but it is necessary to conclude strictly verifiable agreements prohibiting the creation and application of any kinds of space systems. But in the analysis of the future development of ways of armed struggle, it is also necessary to consider that new technical ideas do not cease to appear even if they are not accepted, they go on living, waiting for their time to come.

2 MAIN TRENDS OF DEVELOPMENT WITHIN CONVENTIONAL WEAPONS

Even during the time when nuclear war was considered possible, it was clear that nuclear states will not rush to use nuclear weapons in all theaters of war or against all enemies. The nuclear weapon was designed for use in major theaters of war. In other theaters of war and directly on the battlefield during hostilities, combat tasks should be undertaken using conventional weapons. Apart from missiles, some types of aircraft, artillery, surface ships and submarines have been adapted for use with nuclear as well as conventional munitions.

Therefore, despite the existence of nuclear weapons, conventional weapons have continued to develop at a high pace. And when the practical irrationality and impossibility of the application of nuclear weapons finally became obvious, it actually started a new round in the conventional arms race and in new stages of their qualitative modernization.

In the development of conventional arms, especially of missiles and artillery, the main emphasis of the air and naval forces is on designing precision weapons, which in their effectiveness of application come nearer to nuclear weapons; on the perfection of means of reconnaissance by reconnaissance-strike complexes; on the detection of targets, and guidance systems, automated weapons and troop control facilities. Thus, under the term 'precision weapons' we understand: reconnaissance-strike and reconnaissance-fire complexes, automated fire control systems, anti-tank rocket complexes, homing artillery shells, airforce missiles and missile forces, naval forces, anti-submarine rockets, guided air bombs and cassettes. Great importance is attached to radioelectronic warfare, as being able to paralyze enemy control and the use of major weapon types, and to increase the effectiveness of friendly weapons.

Development of Aircraft and Other Airborne Vehicles

The ever-increasing role of the airforce in the execution both of strategic and operational-tactical tasks is connected with the absence of medium-range missiles, and the appearance on the battlefield of large numbers of high-mobility targets, which are easier to detect and hit from the air. In the next 15 to 20 years a 1.5–2 times increase in the range of detection and launching of missiles against enemy airborne and ground targets is expected thanks to centralized direction and control of aircraft by AWACS (Airborne Warning and Control Systems). Higher levels of aircraft protection will be achieved through the application of 'stealth' technology, the armoring of helicopters, and the use of other new means of active and passive protection.

Speaking of long-range strategic aviation, at present the Russian and Ukrainian strategic air forces are based on: long-range bomber TU-22 MZ, strategic missile-carrying aircraft TU-95 MS, strategic supersonic bomber TU-160, and refueling aircraft IL-78 T and ZM. The flightspeed of TU-160 is more than 2,000 km/h. The main US strategic bomber B-1B has a maximum speed of 1,700 km/h, a flight range of up to 16,000 km and a similar flight ceiling. With the help of tankers like KC-10 or KC-135 such planes can reach practically any point on earth. In August 1993 at the 'Global Enterprise' exercise, two B-1B aircraft for the first time made a round-the-world flight in 45 hours, with one in-flight refueling and one short-term landing on Diego-Garcia airbase in the Indian Ocean. New versions of these aircraft will make it possible to perform such long-distance flights without intermediate landings.

By the end of the century the USA plans to have 210 strategic bombers, including the newest B-2A versions, made with 'stealth' technology.

Rocket aircraft, combining aerodynamic and ballistic principles of flight, will be further developed. They can fly in dense layers of atmosphere like ordinary aircraft, and also become artificial satellites around the earth when they reach orbit.

The speed, ceiling and flight range of aerodynamic flying vehicles (aircraft, helicopters and unmanned vehicles) is increasing. Turbojet engines make it possible to reach flight speeds (even in the atmosphere) corresponding to Mach 5, and in the future with the help of ramjet engines, the flight speed could be brought up to Mach 7 to 8. A transition to rocket engines with the achievement of high rocket speeds is also possible. But aerial vehicles like these will be designed on principles of hypersonic aerodynamics.

The practical ceiling of supersonic aerodynamic aircraft already reaches 25–30 km. But according to the opinions of some designers, in the long term the practical ceiling of aircraft can be raised to 80–100 km. At the same time, especially in the development of tactical aviation, an ever-growing importance is attributed not only to an increase of flight altitude, but also to the ability to fly at extremely low altitudes to complicate detection by radar-tracking stations. With the same view in mind, 'stealth' aircraft are being improved.

In order to increase the operational range of aircraft, their aerodynamic qualities are being improved, and work is being carried out to increase onboard fuel capacity and reduce its consumption.

It is expected that in future the installation of nuclear-power plants on aerial vehicles could help to solve the problem of the increase of operational range of aircraft.

Taking into account that under modern conditions it becomes more and more difficult to ensure the survival of airfields with large concrete landing strips, VTOL (vertical take-off and landing) aircraft are being intensively developed to allow a more dispersed and secretive basing of aircraft.

Armament, navigation methods, aircraft control and combat helicopters are also being improved. The greatest importance is given to the increase of the range of detection of targets, and to missile launches against air, ground and naval targets, and also to accuracy of aim and fire. The main point is to launch missiles and drop bombs, without entering the zones of effectiveness of enemy air defense means.

Tempting and realistic prospects have opened up in the use of airships – piloted air ships, which are lighter than air. They can be equipped with diesel, turboprop, and for the future, even nuclear engines. Even now, a prototype of airship 'Sentinel-5000' can fly continuously for two to three days, and, with refueling, for up to 30 days.

In the future, their flight duration could be considerably increased. The most optimal application of airships might be to use them as platforms for the installation of early-radartracking detection systems, providing the continuous supervision of movements on ground, under water, and of aerial targets (including low-flying cruise missiles). Initially, airships could be equipped to carry radartracking detection systems, currently installed on E-2C 'Hawk-eye', E-3A 'Sentry', and AWACS aircraft. In the future, brand new, lighter radar-tracking stations with a phased array can be created which, deployed at the altitude of up to 4 to 5 km and higher, will have a considerably larger (2 to 3 times) range of detection than ground-based radartracking stations.

Radar systems deployed on airships will be of special practical value for naval forces. They could illuminate significant sea and air spaces; provide earlier detection of air and naval targets; radar cover and support of troop control, especially when escorting aircraft-carrier and other naval groupings in the oceans; guard naval and air bases; perform meteorological tasks and so on. Radar and other reconnaissance means deployed on airships could be used to support mobile forces, in military conflicts in distant regions where stationary and other systems of radio support are not available.

The idea of utilizing airships to establish radar systems of surveillance, and the early detection of aerial targets and missiles, becomes especially acute for Russia, the CIS countries, the Baltic and other states – which as a result of the destruction of the unified air defense system that existed in the USSR are compelled one way or another to organize a system of radar support anew. From this perspective it would be more cost-efficient to start to create a future-oriented, new system of air defense, and to make a qualitative jump in this direction, without wasting time and resources on the re-creation of obviously out-of-date systems of air defense. Whoever is last in creating any new system, and drops the remnants of the old systems, will in the long term attain the greatest gain, if, of course, he is not losing the chance to solve pressing problems with due foresight.

But the above listed examples do not limit the possibilities of airships. In the future, they can be widely used for the transportation of material stock, heavy equipment and mobile troops to remote regions.

Recently, conflicts have been arising in distant and difficult-to-reach regions. This leads to changes in the ways aviation is used in the beginning of war, and necessitates an increase in the operational radius of tactical aircraft, the possibility of inflight refueling, and greater secrecy of flights. NATO's airforce has been surpassing Soviet aviation in these respects. But now, with the formation of a buffer zone for Russian aircraft of not only the third, but also the fourth generation, combat opportunities are largely reduced.

For example, a tactical fighter like F-117A, with a small radar signature and able to fly at low as well as high altitudes, can cross a buffer zone undetected by air defenses (before the approach at a distance of 20–30 km). B-2 bombers will be even better in this respect. The bomb payload of this aircraft is 18 tons, allowing it to carry 16 missiles of the 8-RAM-P type, with a tactical radius (without refueling) of up to 10,000 km. Onboard equipment of the F-117B and B-2 aircraft enables them to detect the operation of air defense radar stations long before they could possibly be detected themselves, and automatically to adjust the route. Targeting equipment is suited for operations at night and in adverse weather conditions. The combat performance of the F-117A has proved true in the war of coalition forces against Iraq. As far as the B-2 is concerned, estimations by American experts state that a pair of B-2 bombers, without support and refueling aircraft, might have carried out the same mission which took a formation composed of 32 strike aircraft, 16 fighters, 12 air defense suppression aircraft and 15 refueling aircraft in the war against Iraq.[6] The YF-22 and YF-23 fighters will have an even better performance. Among less noticeable means of air assault, the ground, air and sea-based cruise missiles acquire special importance. For example, new versions of the sea-based missile ACM-109S 'Tomahawk', flying at altitudes as low as 100 m (with an effective reflecting surface 0.05 m), can be used for long-distance flights of over 1,600 km with accuracy of up to 30 m.[7]

Air Defense Facilities

Enhancement of combat capabilities for the airforce, with increased numbers of other aerial targets, sets new requirements for air defense facilities that will enable air defense to repulse massive strikes of hostile aviation and tactical missiles. Thus, the most urgent problem is to further improve the already exhausted resources of detection, tracking and target selection equipment, and active guidance methods against aerial and outer space targets. The problem is even more complicated by the creation of 'stealth' aircraft and by the increasing use of low-flying cruise missiles. Therefore, new ways of developing radar systems are emerging, based on a backscatter principle, which receive signals from ionized atmospheric layers, from ground or water surfaces, from gaseous missile jet exhausts, the radio emission of jet-engines, from low frequency soundwaves that accompany missile launch, and in a number of other directions. The most promising application of lasers is the location and identification of targets, with the positioning and guidance of active air defense means. As has already been said, the installation of radar stations and television systems on airships and other kinds of aerial platforms, could be rather useful.

The development of active means of destruction of air targets in the form of universal homing systems, capable of hitting various targets, is also envisaged and is already implemented in the latest versions of such systems as 'Patriot' (USA) and S-300 (Russia). For example, the air target destruction effectiveness of the 'Patriot' complex is 12 times higher than that of the 'Hawk-Hercules' complex, and six times higher than that of an advanced version of 'Hawk'.

During the war with Iraq in 1991 'Patriot' successfully forced down not only aircraft, but also R-300 Scud ballistic missiles. Russian anti-aircraft missile systems S-300A, 'BUK' and 'TOR' are able to simultaneously follow and attack dozens of aerial targets, flying at a speed of 10,000 km/h and faster. But the new army air defense methods should be capable of destroying missiles flying at speeds considerably higher than that of R-300 missiles. The struggle against ballistic missiles will be conducted by means of antimissile defense.

In order to ensure the greater effectiveness and survival of air defense facilities, not only army systems, but also electromagnetic and anti-aircraft missile systems deployed deep inside the country will become mobile, and probably only a few of them will remain stationary.

Naval Forces

The exhaustion of Earth's resources will sooner or later compel mankind to turn to the rich sources under the seas and oceans. To provide safe access to these resources, one would need powerful naval forces. Therefore, the leading states of the world will give priority to the development of surface and submarine forces and amphibious means, and to the perfection of centralized monitoring systems of naval situations, and the control of fleet and aviation forces. In the naval forces of the leading states of the world, submarine forces (antisubmarine and nuclear missiles), as well as naval missile-carrying aviation have become the major fleet forces. Nuclear submarines were up till now adapted mainly for the employment of nuclear weapons. Ballistic missile submarines did not carry conventional weapons, except as a means of self-defense. Only in the case of a significant reduction or complete elimination of strategic nuclear forces will the question of their re-equipment with conventional ammunitions arise. It is possible that the displacement of submarines will increase up to 10,000 tons and more, with diving depth of more than 1,000 meters.

The USA surface fleet has aircraft carriers acting as mobile air bases, with up to 100 and more combat aircraft and helicopters on board, as well as antisubmarine ships (destroyers, frigates and escort ships) equipped with sonars, homing torpedoes, guided rocket-thrown torpedoes, jet bomb throwers for launching depth bombs, and antisubmarine weapons. Britain and France both have attack or antisubmarine aircraft carriers. New attack aircraft carriers with more powerful aircraft park and missile systems are developed. The deployment of V/STOL (vertical/short take-off and landing) aircraft will increase the number of aircraft on aircraft carriers up to 150–180. At the same time there are no plans to construct any more anti-submarine aircraft carriers in the future.

As far as Russia is concerned, it has several aircraft-carrying ships with aircraft of limited operational radius (not more than 150–200 km), left from the former USSR. In the forthcoming decades Russia will hardly be able to build any more attack aircraft carriers.

In the naval forces of the leading states an important role will continue to be played by heavy and light cruisers, including nuclear cruisers. Assault and mine warfare ships will be further developed and technically perfected. Air defense systems for ships will be further enhanced.

The increased role of mobile forces raises the importance of such components as assault and marine forces. The construction of a large series (up to 30 units) of assault ships under the new project LX, capable of taking 700 marines, several helicopters, and one or two aircushion boats, with an overall surface area for armament and military equipment of up to 2,320 sq. m. is now being actively considered.

For the USA, naval assault ships, along with transport aircraft, have become a major material basis for the realization of military policy and strategy, connected with rapid response to military conflicts and maintenance of readiness for simultaneous participation in two local wars.

In the construction of new ships an aircushion principle is now widely used, and *ekranoplans* (wing-in-ground) and hydrofoil boats are being developed. Fundamentally new power plants for large surface ships and submarines are being designed. The practical applicability of some new ideas is being studied, like the design of a boat able to fly over water at the speed of aircraft; or the building of partially submerged ships – underwater tankers, assault ships, submersible cruisers and aircraft carriers – ships with large underwater hulls and a small superstructure, to ensure the high survival of ships and the secrecy of their operations. Special attention is being paid to the increase of power, range and accuracy of application of armaments on ships of all classes. Sonar equipment, reconnaissance and radio-electronic warfare facilities and automated control systems of naval forces are also being improved. Ways are being sought to ensure lower noise levels of submarines, while the possibility of finding biological and chemical structural changes in water along the submarine trace, are under investigation.

The Ground Forces

Within the Army, major attention will be directed towards the perfection of missiles, artillery, tanks, air defense facilities, electronic warfare, reconnaissance, and troops and weapons control. After the conclusion of agreements on the elimination of missiles with an operating range of more than 500 km, tactical missiles have been improved into multipurpose missiles with an increased range, capacity and accuracy of munitions. The Soviet 'Oka' missile with a range not exceeding 500 km was the most advanced in this respect. The decision to eliminate these missiles was taken by the then Soviet authorities purely for political reasons. But such solid fuel missiles of high accuracy (within the ranges stipulated by international agreements) remain very promising. And not all states have joined in agreements on the elimination of intermediate and short range missiles, and can still operate them. The use of lighter, heat-resistant materials will lessen missile weight and dimensions, enabling helicopters to transport them quickly to the launch point. The utilization of various homing systems will result in higher accuracy.

With the artillery, further perfection of rocket-assisted projectiles, with high-explosive at the forepoint and a rocket engine at the base, provides a doubly increased firing range while at the same time reducing the weight of the artillery piece. Self-propelled armored artillery will be extensively used, owing to its enhanced survivability. Previously, the explosion of shells on an artillery battery position usually led to hitting unprotected firing crews with shell fragments, resulting in the destruction of the battery. Now, only a direct hit can cause the destruction of self-propelled armored artillery, and this changes the standards set for artillery assigned for counter-battery fire.

Automatic mortars (with a rate of fire of 25–30 rounds per minute), and volley fire from rocket artillery appear to be rather effective against infantry in trenches, and against large group targets. An increase in the firing range of RS30 (rocket artillery system of volley fire) from 40–70 up to 100–120 km is expected.

In Russia this system is represented by the 'Smerch' complex, in the USA by MLRS (multiple launch rocket system). The latter can be equipped with anti-tank submunitions, with homing war heads in final trajectory. The launch range of such rockets is up to 40 km. They can be charged with 28 anti-tank mines. The two-system salvo ensures a 1–1.5 km-long anti-tank mine field. In the Persian Gulf, Americans for the first time used MLRS to launch the tactical missiles ATACMS (army tactical missile system) that replaced 'Lance' missiles. MLRS was the only system of field artillery capable of functioning in automated coordination with the 'Abrams' tanks, 'Bradley' APC and assault aviation, that provided target designation and fire adjustment.

It is also necessary to expect a further perfection of tanks, armored personnel vehicles, anti-tank and air defense facilities with respect to the increase in accuracy and fire power, and the hardening and survivability of weapon systems against hostile fire and radioelectronic warfare. The proportion of army aviation intended for destruction of the enemy, as well as for troop transportation and combat support, will increase. The further perfection of night vision devices in weapons and equipment among all types of troops will increase their capabilities in night combat actions. Universalizing equipment, creating families of vehicles with the same types of parts will facilitate the operational and technical support of troops. (A detailed development of these kinds of weapons will be presented in the review of methods of combat actions.)

More effective means of remote mining, minefield clearance, stream crossing, traversing other obstacles, and entanglements on a battlefield are emerging.

The conduct of combat actions in local wars and conflicts sets specific requirements on arms and equipment. As a result, existing weapon types are modernized and new systems are developed. Israeli 'Merkava' tanks, APCs in South Africa and light aircraft used in South American countries to combat drug dealers are examples. Experience from combat actions in the Kabul region has shown that well-protected helicopters equipped with the means of detection of hostile fire are extremely necessary.

If major importance is given to precision weapons systems, then it will be impossible to solve local wars without them. In order to prevent casualties among the civilian population, not large areas of destruction but selectivity and accuracy in hitting hostile fire is needed. Mortars, ATGMs (anti-tank guided missiles), combat helicopters and assault aircraft would be most effective in these conditions, being able to operate directly on the battlefield.

Of particular difficulty in local wars and conflicts is the coordination between the airforce and air defense, with the identification and designation of ground troops operating not only in front but also in the rear and on enemy flanks. There were several cases of destruction of friendly troops by American and British aircraft during the Persian Gulf hostilities. This is why more advanced identification capabilities are required.

More effective means of close combat are sought for police forces and other special troops (silent small arms, grenade launchers with ammunitions of enhanced capacity, dazzle and gas weapons, body armor vests and so on). Mobile troops will also be equipped with similar means.

Small arms are being modernized and will continue to be improved, although F. Engels noted in the last century with respect to small arms that 'in this direction the era of development in essential features is completed'.[8]

In the perfection of small arms the main emphasis is on increase in reliability, rate of fire, and weight reduction.

Large-caliber machine guns (mainly 14.5, 23 and 37 mm) are widely adapted among army troops, airforce, air defense and naval forces. Silencer devices are being fitted on weapons. The range of night vision devices is increasing. A trend towards extending such weapons is revealing itself. For example rifles (carbine) can be used as anti-tank means, for which purpose the barrel is mounted with a grenade launcher for the firing of hollow charge grenades.

Promising grenade launchers such as RPG-2á and RPG-29 are capable of piercing the armor of any tank. Individual weapons having two barrels firing a grenade and rifle bullet, have been developed.

Various anti-tank means are: long-range rocket propelled anti-tank rifles; anti-tank rocket-assisted projectiles guided by radio, cable and homing devices; recoilless anti-tank guns; improved rifle and smoothbore guns of field artillery and tank guns that use hypervelocity armor-piercing projectiles, plus tandem and hollow-charge projectiles. Major attention is being paid to the increased range and armor piercing capacity of anti-tank systems. Semi-active, infra-red self-guidance systems and homing and laser guidance systems are being developed to increase accuracy of fire.

The infantry, tank units and other services in the armed forces are being equipped with new portable air defense facilities such as 'Strela', 'Stinger' and others but with higher performance in respect to height and range.

In the future, the mobility of all types of troops will be enhanced. With this aim, greater attention will be given to the perfection of heavy weapons-carrying vehicles and transport means with respect to their increase in speed, their efficiency as multifuel engines, and their crosscountry capability in adverse terrain conditions such as marshes, sand or deep snow. Wheeled vehicles with several leading axes on rollers or tracks are being developed. Taking into account the rather limited endurance of tracked vehicles, a basic technical idea used by Soviet tanks of the 'BT' types has not lost its importance; 'BT' tanks could change their running gear, changing if necessary from wheels to tracks. In building new generations of vehicles the principles of 'zero pressure' and 'aircushion' are used. All basic combat and some transport means will become amphibian, to accelerate water crossings.

For all armed forces, outer space support has become one of the major ways of combat support. Space systems facilitate navigation on the ground, the air and on the sea, and will help to detect hostile objects and targets and transfer data in real time, thus assisting to a more accurate application of weapons.

3 ARMS REDUCTION AND ITS INFLUENCE ON THE NATURE OF ARMED STRUGGLE

The reduction in nuclear and conventional arms will be accompanied by their qualitative perfection. A statement made by the secretary of state for defense of Great Britain that the British Armed Forces will be reduced and that the reduction will be compensated by new quality, reflects the general tendency among western countries. At the same time Russia, the CIS and other east-European countries are seeing the destructive conversion of their defense industries, with rigid restrictions in the delivery of new technologies, and so cannot survive competition in the qualitative perfection of weapons, and are destined to stay significantly behind. In the next ten to 15 years China, India and many countries of the Third World will not be able to reach new technological levels in the development of weapons systems. In these cases armed struggle will less often be conducted between opponents having about equal military technological opportunities, as for example in the Second World War, the Arab–Israeli wars, the Iran–Iraq war, and in other cases.

In the foreseeable future, especially in local wars and conflicts, armies and naval forces with the newest military equipment might oppose forces having only outdated weapons. Even in the Gulf War the confrontation was, as the saying goes, between wars of two epochs: war with the application of ultra-modern technology led by multinational forces; and war using outdated weapons on the part of Iraq.

Arms reduction produces its influence in the structure of armed forces and on some established strategic categories. At present and in the given situation, the upkeep of nuclear parity and the opportunity of retaliation have lost their former importance. In general, it is possible to say that nuclear weapons will act in the future as a means of strategic deterrence. It is sometimes believed that the main point of strategic stability is not the quantitative and qualitative ratio of nuclear arms, and not even parity, but only the availability of nuclear weapons for the parties concerned (even in minimum quantities), and the degree of apprehension of the consequences of their application. This is hard to agree with, for whichever party possesses superiority may deliver a sudden disarming strike and avoid the threat of retaliation, by applying systems causing radioactive contamination, or just by using conventional precision weapons. Therefore, if not actual parity, then at least there must exist sufficient means to ensure retaliation, capable of inflicting significant damage upon the opposing party. Besides, under conditions of the ongoing proliferation of nuclear weapons, there is the possibility that it might be used illegally, or for nuclear blackmail, or by terrorists.

The moment has come when arms reduction negotiations need to take into account not only quantitative, but also the qualitative parameters of weapons, including systems of long-range detection, control and support of their combat application. For example, depriving Russia of radar early warning stations in Krasnoyarsk, Riga, Mukachev and Mingechaur gives the advantage to the US, which has retained early warning and detection means such as AWACS and 'Sosus' capable of 'seeing' at distances of hundreds of kilometers. For the USA this also means the control of aircraft and submarines in ocean water zones, and the accurate application of air and naval strategic forces, even under formal agreements concerning the quantity of aircraft or submarines. Geostrategic positions of states should also be taken into account. The fact that Russia remains a continental, and the USA an oceanic maritime state, cannot but influence the structure of armed forces, including strategic nuclear forces. For example, by losing air refuelers to the Ukraine, Russian strategic aviation in many respects has lost its operational effectiveness. Is there any reason under such conditions to only calculate the number of platforms?

Besides, many of the questions of strategic and conventional arms reduction were until now discussed apart from military and doctrinal and even operative strategic thinking. At the same time 'the sufficiency of defense' and particularly of strategic offensive methods, is in many respects determined not only by the quantity and quality of arms, but also by ways of their combat application. If, let us say, some countries will not abstain from the principle of first use of nuclear weapons, then others during further reduction of strategic offensive methods will simply not be capable of adequate retaliation, so that their strategic nuclear forces will actually lose the ability to reliably deter a potential aggressor. All this creates completely new conditions for solving strategic problems.

One such question concerns the organization of strategic non-nuclear forces. The elimination of medium and short-range missiles, of basic types of tactical nuclear weapons and strategic arms, brings into being new conditions for the maintenance of strategic stability in the world.

However, after START-1 and START-2 the subsequent agreements signed by the USA and Russia on the significant reduction of nuclear arms, can actually be effected only together with China, Great Britain, France and other nuclear states.

With a view to reducing nuclear danger alongside the reduction of the quantity of nuclear weapons it might be expedient to accomplish mutually-coordinated measures on the removal from combat duty of a certain part of the nuclear forces; to change the targets and planning order in the application of nuclear weapons; to alter the tasks of strategic nuclear forces, their structures, and all combat systems.

The Paris agreements also established levels of conventional arms for NATO and the Warsaw Treaty Organization. But after the dissolution of the Warsaw Treaty Organization and the disintegration of the USSR, the ratio of armament, norms of contents, and the regions of their concentration have significantly changed. This necessitates the conclusion of new international agreements in Europe, Asia and other regions, in view of the presently forming multipolar world, and the interests of different states. The basic parameters, defining the directions of such agreements, could be given by the UN, and the specific levels of arms could be decided by bilateral and multilateral agreements.

One of the important measures in the international arena should be the exclusion of the overwhelming superiority of any single state or group of leading states, in conducting a conventional war with the application of new technologies. Especially dangerous is the capacity of certain states to deliver powerful fire strikes on the state, economic and military objects of other countries from a distance, even before the clash of ground forces. If in such cases it is feasible to oppose the aggressor with an adequate threat of retaliatory strike against his territory, strategic stability can be breached. Therefore, whether before or after the sharp reduction of nuclear arms, especially with non-nuclear major powers, the question of the creation of strategic non-nuclear forces, capable of at least threatening with destruction the major objects of strategic importance in the confronting party, will definitely emerge.

Some military experts emphasize that these forces will be some sort of non-nuclear deterrence forces, whose main purpose will not be so much combat application, as the creation of the potential threat of their utilization in action with serious consequences for the aggressor.[9] The peace loving states might undertake such actions only in exceptional cases and in retaliatory operations. But the need for the creation of non-nuclear strategic forces is becoming urgent.

In solving this task there will be complex problems, connected first of all with the separation of nuclear and non-nuclear means, but problems, as the saying goes, exist in order to be solved. Therefore, operational strategic, scientific and technological research in this direction will go on. For instance, strategic non-nuclear forces could be deployed in specially stipulated regions.

The basis of strategic non-nuclear forces could be constituted by strategic aviation without nuclear components, then sections of missile and submarine forces dispossessed of nuclear ammunitions.

Certainly, the conception of strategic non-nuclear forces can open up a new direction in the arms race in the field of non-nuclear technologies, but in due course it will provide the great powers with an easier way to discard nuclear weapons.

Such are some of the prospects in the development of weapon systems and military equipment, that will affect the nature of armed struggle.

NOTES

1. Report of the Secretary of Defense to the President and Congress, p. 25.
2. *Segodnya* (3 August 1993).
3. *Za rubezhom*, No. 12 (1994).
4. *Kosmicheskoye oruzhie: dilemma bezopasnosti* (Moscow, 1989) pp. 81–91.
5. Cited from: A.S. Orlov, *V poiskakh absolutnogo oruzhiya* (1989) p. 254.
6. *Voyennyy vestnik*, No. 15 (1992) p. 3.
7. Ibid.
8. K. Marx and F. Engels, *Sochineniya, Izd. 2-e*, Vol. 20, p. 174.
9. *Voyennaya mysl' (Military thought)* No. 1 (1992) p. 49.

CHAPTER 3

DISTINCTIVE FEATURES OF FUTURE ARMED STRUGGLE

Best of all is to defeat a hostile army without a fight, or else to deprive it of the possibilities of resistance.

Sun-Tsu

These soldiers are bravehearts: they always climb where the wall is thickest.

Admiral de Robeck

The entire art of war consists in the execution of a well-grounded and thought-out defense, with the subsequent transition into a swift and resolute offense.

Napoleon

1 NUCLEAR WEAPONS AND ARMED STRUGGLE

How the Strategy of Global Nuclear War Was Born and Has Become Obsolete

The political and military–technical factors spoken of earlier, predetermine sources of confrontation and can also secure the most resolute political ends. Yet until recently any possible world war was viewed as an adamant collision between two opposing systems – capitalism and socialism. It was supposed that given a practically unlimited attainability in nuclear weapons able to destroy any hostile objects in the short term, this war in strategic respects would turn into a global confrontation of the two largest coalition groupings of armed forces, unprecedented in history in its scope and brutality. On the other hand, it was clear that new weapons put certain limits on political and strategic aspirations. But this circumstance was not always fully taken into account, and it seemed that a world war would inevitably be nuclear. A theory of nuclear war originated, which thoroughly investigated the combat properties of nuclear weapon missiles, the ways of their possible application, and protection measures for troops and populations against weapons of mass-destruction.

Some theorists obviously exaggerated the potential consequences of even limited application. Thus, some scientists wrote that dropping 180 nuclear bombs on the USSR, as was foreseen under the 'Dropshot' plan, would result in heavy contamination by radioactive deposits, such as radionuclides, not only on the territory of the USSR, but also in the USA, leading therefore to the likely destruction of both countries. However, since World War II, 512 nuclear explosions have been produced in the earth's atmospheric layers, and the consequences are known. Other theorists, on the contrary, underestimated the consequences of the use of nuclear weapons.

Within theories of nuclear war completely unrealistic measures were presupposed: for example, it was conceived that after an exchange of massive nuclear strikes, the restoration of the operational effectiveness of troops and navies, with continuation of combat actions, and a rate of march on the offensive of up to 100 km per day would be possible. During combat preparation, this most complicated problem was often simplified in the most superficial way and led to funny things. Marshal of the Soviet Union S.S. Birusov wrote about one of them. During tactical training an officer was given an exercise in which his unit on the march was attacked with nuclear weapons. He quickly made a decision and issued a command: 'Continue the march on the specified route. Speed up movement.'[1] Despite similar instances, during the period when opposing parties had a comparatively limited amount of nuclear weapons, and the first plans for nuclear war ('Dropshot' initial variants of 'SIOP' [single integrated operation plan], and Soviet plans for the application of strategic nuclear forces) were being developed, it was possible at that stage to admit that under conditions of the use of nuclear weapons, organized military actions, though with heavy losses, are still possible.

This was confirmed during experimental exercises with the actual application of nuclear weapons in the USA and USSR. For instance, during the Totskoe exercises of September 1954, where I happened to participate, after the explosion of a nuclear bomb of 20 kt yield, the attacking troops who wore protective gear both inside their tanks and outside, subsequently crossed the nuclear explosion area and received a dose of radiation equal to 0.02–0.03 roentgens (4–5 times less for those inside tanks). But subsequent grave consequences of ionizing radiation were revealed. One theoretical work reflecting on the real possibility of running combat actions under the employment of nuclear weapons was O. Mikshe's book *Nuclear Weapons and the Army* published in 1955. Special manuals for combat operations with the application of nuclear weapons were developed for the NATO and Warsaw Pact armies.

Nuclear questions were later also reflected in the combat regulations of many countries. After the intervention of China in the Korean war, the American government was ready to start a nuclear war, and only the intervention of the prime minister of Great Britain, C. Attlee, could keep President H. Truman from using the nuclear bomb in Korea.

During the Berlin crisis in 1961, US President J. Kennedy declared in a TV address that the United States, in the struggle for the status of Berlin, should make a choice between nuclear war, and war with conventional weapons. Kennedy chose the latter, which was indicated by the drafting of about a million reservists.[2]

Diametrically opposed judgments have been expressed in the western press on matters of nuclear war. For example H. Kahn, proving the acceptability of the thermonuclear war as a political instrument, asserted that 'war is awful, but so is the world. Mankind will survive a third world war with the application of nuclear missile weapons. Colossal destruction and numbers of victims will not eliminate normal life for survivors and their descendants.'[3] B. Goldwater has spoken even more definitely in the same spirit, against the fear of nuclear war. He wrote: 'We are strong, we shall win, we should not be frightened by victims and ruins, the main thing is to defeat communism.'[4] In the document entitled 'Directive instructions in the field of defense', which contained the basic military and political goals of the USA for the years 1984–88, it was emphasized that the USA should get ready to conduct protracted nuclear war with the USSR: 'The United States should gain the victory in this war and compel the Soviet Union to search for the prompt termination of war on conditions favorable for the USA.'

The American expert on nuclear weapons E. Knoll, characterizing an afterwar situation, has come to the conclusion that:

> The United States then had the monopoly on the most effective weapon ever invented by human genius. Armed with such a weapon, this nation could carve anew the postwar world at its own discretion. Besides, another great superpower triumphant in the Second World War – the Soviet Union – could be kept in check by American nuclear monopoly ... The new role of nuclear weapons – to keep and to expand leadership of America in the world ... In the highest deliberation of the Truman administration and foreign policy it was the beginning of establishing the domination of the United States' interests all over the world.[5]

Another American L. Pauling – winner of the Nobel Prize for Chemistry and the Nobel Prize for Peace – and a passionate champion of the prohibition of nuclear tests wrote: 'I am convinced that the creation of this terrible weapon compels us to begin a period of peaceful history, a period of peace and common sense.'[6]

P. Krosser's book *Dialectics of military material and its consequences* was devoted to proving one idea: 'military equipment which used to serve as the means of allowing the military to win wars, as in the pre-nuclear period, is in the nuclear age becoming a means of destruction of all participants of nuclear-missile war.'[7] Similar views were also shared by J. Bernal.[8]

The official point of view of the Soviet political and military leadership was expressed in 1971 by the USSR defense minister Marshal A. Grechko: 'The modern defensive power of the USSR and other socialist countries gives us reason to declare that we are in the possession of armed force of such colossal capacity that it will break any aggressor.'[9]

It should be said that the appearance in the Soviet Army of the first manuals on combat actions under conditions of nuclear weapon application, hastily copied from American manuals, were met by some military leaders and many officers with latent, and sometimes even open opposition. The commander of the Belorussian military district Marshal S.K. Timoshenko, even ordered that these manuals be not shown to anybody. And only after pressing demands from the defense minister were these manuals sent to troops of the district.

Marshal Timoshenko did not believe that nuclear weapons could be utilized, and expressed the conviction that they would experience the same fate as chemical weapons of the Second World War. Most officers with combat experience were disturbed by the fact that hopes to solve all basic combat problems with the help of nuclear weapons would result in the degradation of military art. These officers were then mercilessly criticized as conservatives. but experienced people, even without broad education, have a special vision which often proves to be true in life.

Some military theorists are at this time distressed by the fact that military art insufficiently takes into account the influence of the application of nuclear weapons in theconduct of armed struggle. But there is no sense in adapting military art just to the weapons, which then cannot even be applied.

However, despite the continual threat of nuclear weapons, as thousands of nuclear ammunitions and ways of their delivery are accumulated, it has become more and more obvious (though it was not understood by everyone at once) that a war with massive application of nuclear weapons is unfeasible, and there will be no winners in it.

Be as it may, even after the highest-ranking political and military officials came to a conclusion on the impossibility of winning a nuclear war, this weapon, hanging over humankind like a sword of Damocles, continues to render tremendous influences on policy and military strategy, by giving them certain reliance on it, as well as a restraining fear. For example, if it had not been for the threat of the utilization of nuclear weapons, it might not have been possible to overcome the Caribbean Crisis of 1962 by way of negotiations. The dangerous nature of modern warfare constrains to a certain degree the deployment not only of nuclear, but also of conventional weapons. For example, according to the report of the Brooklyn Institute 'Use of US Armed Forces as an instrument of foreign policy in 1946–1975', the USA (except for wars in Korea and Vietnam) has used its armed forces for protection of national interests in 215 cases. And only in 18 incidents (and outside the frameworks of the East–West confrontation) did they open fire. In other cases, the use of military force was expressed only in a military presence or a demonstration of force.[10]

Great prudence was displayed in other spheres of activity as well. It certainly did not mean a refusal from political and strategic purposes in themselves, as new ways to achieve them were sought.

First, some leading states developed methods for putting political and economic pressure on other states, interfering with the realization of their interests, or on ways to undermine them from within, thus accomplishing political goals without using military force.

Second, new methods of waging local wars were sought. In the Cold War period NATO and the Warsaw bloc, lead by the two superpowers, struggled to restrict each other, and to expand their spheres of influence by local wars in Korea, Vietnam, the Middle East, Afghanistan and other places.

As early as the beginning of the 1960s when the USSR procured the possibility of missile strikes not only on western Europe but also on the USA, many political figures and military theorists raised the question of the total rejection of war and a shift to limited local wars.[11]

B.H. Liddell Hart, for example, wrote that none of the fighting sides would ever be able to gain victory in total nuclear war, and called for the rejection of such war. In order to resolve the defense problems of the West, and for the successful conduct of limited wars, he recommended powerful conventional non-nuclear forces, the application of which do not create dangers like those of total nuclear war.

R. Osgood devoted his book 'to the general theory' of limited war, proving that such a war is the most reasonable political decision for the USA. He and E.J. Kingston-MacLory assumed the possible application of tactical nuclear weapons in local wars. M. Taylor proved the need of replacing the strategy of 'massive retribution' with the strategy of 'flexible response'. Wrote Taylor: 'We have lost our former nuclear monopoly ... The leading strategic concept of massive retaliation has reached an impasse ... The necessity has arisen to reassess our strategic requirements.'[12]

Originating in the USA in 1961, the 'strategy of flexible response' was accepted in December 1967 as the official doctrine of NATO. This strategy was directed not only at global nuclear war, but also stipulated a wide range of war and military conflicts without the application of nuclear weapons. In this connection, a course towards the substantial growth of general-purpose forces was taken, and in the period from 1961 to 1968 the aggregate number of the US armed forces increased from 2.5 to 3.5 million servicemen, and the number of divisions in the Army and Marine Corps from 14 to 24.

Even in other nuclear states, conventional weapons made up about 80 per cent of all military expenditures, and not less than 90 per cent of personnel of the armed forces were assigned conventional units.

Orientation of Military Doctrines of the Great Powers

Alongside orientation on local wars, NATO countries did not completely dispense with preparation for world war. In a report by the US Army research and development chief General Gaywin, it was put forward that local wars are only a way of preparing for total war, that the ultimate goal of US military policy is general world war, and he emphasized that 'without it the theory of limited war has little sense'.[13]

New inventions (such as multiple warheads of high accuracy, cruise missiles, outer space systems, advanced means of reconnaissance and control) all show that American policy had once again taken the course of nuclear war. There were expectations of a disarming strike, depriving the opponent of the possibility of retaliation, and mainly of course to involve the USSR in a new round of the arms race, with the aim of economically exhausting it. It is one of the evident examples of how new weapons influence conceptions of policy.

USA policy in the 1980s tightened (with the coming to power of R. Reagan), and within the frameworks of the 'flexible response strategy' a landmark was taken again not only on conventional, but also on general nuclear war, with the aim not simply to 'restrain' or 'impose retribution', but to achieve victory over the 'evil empire'. In accordance with these directions, priorities of military construction were defined. Without refusing the build-up of conventional weapons, the main emphasis was made on strengthening nuclear potential as a means of delivering a first disarming strike, with the simultaneous creation of antimissile defense using elements of space-based systems; and in conventional weapons – the creation of new kinds of precision weapon systems, with an efficiency 5–6 times higher. As a whole, Reagan's orientation on 'horizontal (geographical) escalation' meant a readiness to enter battle with communism at any point on the globe, and to participate in 2.5 wars, but, in fact, in as many wars or local conflicts as the forces permitted.

In the USSR the highest political and military leader, defense minister Marshal R. Malinovskiy, personally maintained that local wars would inevitably develop into world war, and that primary emphasis was made on preparing for such a war.

This was reflected in fundamental political and military documents. These works included the books, *Problems of the Revolution in Military Affairs*, which contained a chapter by Marshal R. Malinovskiy, and the three editions of *Military Strategy*, edited by the Chief of the General Staff, Marshal V.D. Sokolovsky. The importance of local wars was underestimated in the USSR, although their tendency to increase was already clearly revealing itself. For instance, before World War II, in the whole world there occurred on average 2.6 local wars or conflicts a year; from 1917 till 1939 3.4; after World War II 15.6.

But gradually the USSR came to realize the necessity of preparing not only for nuclear, but also for conventional war. In the 1960s and 1970s, the quantity and quality of conventional weapons increased. Special emphasis was made on the development of aviation and operative-tactical missiles in the army, airforce and navy. In 1977, after the assignment of Marshal N.V. Ogarkov as chief of general staff, the program of the partial reorganization of the Soviet Armed Forces was developed. Personnel in the defense ministry and general staff were reduced; while their organizational structure was simultaneously improved.

The system of air defense was radically reconstructed. The question of abolishing the air defense service was even raised. An attempt was made to discontinue the construction of expensive aircraft carriers, and expenditures for the requirements of civil defense were significantly reduced. A more advanced system of combat, mobilization-readiness and operational training was introduced. However, many of the plans conceived were not accomplished. And not only because of the inconsistency of defense minister D.F. Ustinov, and fierce resistance among the main commands of the services (it really was like that), but also largely because it is impossible to carry out the reform of armed forces without changing the military policy of the state, and without developing strategic concepts of the preparation and conducting armed struggles appropriate to new conditions.

As before, the main stake was made on preparation for global nuclear war, although after the chief of general staff left his post, Ogarkov was one of the first major military leaders to doubt the 'possibility of warfare with the application of nuclear weapons in general'.[14] For the first time main commands of troops on strategic directions (in a theater of military actions) were created, requiring large additional expenditures which could only be necessary in a world war. Any reduction of expenditures on defense was not presupposed. No account whatsoever was given to the fact that the Cold

War and arms race could not continue any more (especially for the USSR) and that it was necessary to look for other ways out of the resulting impasse.

Nevertheless, despite various fluctuations on separate aspects of the problem, the opposing sides gradually realized that preparation for nuclear and conventional world war could not proceed at the same scale as before. To stop this process, fraught with catastrophic consequences, it was essential to finish the Cold War and, thank God, it was terminated before developing into a real world war. Military policy and strategy, deep-rooted in the ideology of nuclear war, lacking as yet the development of new concepts, applicable under the altered conditions, have reached a most complicated crossroad.

NATO's course on adaptation of the old bloc to new political conditions, is reflected in 'New strategic NATO concepts' adopted in 1991 at the Rome session. It assumes that political approaches to the provision of security will acquire growing importance. Thus, two previously proclaimed principles, those of 'defense' and 'dialogue', are supplemented by a third, 'cooperation'. These principles refer to the former USSR and other east-European countries. The military-strategic aspect of the concept, while leaving intact former principles, stipulates the following:

'Nuclear intimidation', but realization of the 'strategy of flexible reaction' in appropriate circumstances and at a lowered level of nuclear arms;

'Forward defense', reserving it mainly for north-European and south-European theaters of military actions;

'Reduced forward presence', which provides the deployment of major groupings of troops not in proximity to NATO's eastern borders, but in the rear regions.

Under conditions when the 'Soviet military threat' ceases to exist, the Clinton administration considers the main source of threat to the national security of the USA to be: the proliferation of nuclear weapons; state terrorism; and ethnic conflicts growing into long-term centers of military tension.

In the report 'National Security Strategy of the USA' it is emphasized that the

> US Armed Forces play a key role for the achievement of success in the realization of our strategy. Our country has no equals in military possibilities: The United States is the only state able to carry out full-scale and successful operations far from its own borders ... The US Armed Forces should be able to react quickly and effectively, to be ready to fight and to win. To achieve this one needs well-trained personnel, convinced of the righteousness of their course; modern arms, maintained fully operational; realization of the training of

> personnel under close-to combat conditions; sustainment of strategic mobility; creation and support of a reliable system of materiel support ... We also have to maintain sufficient strategic nuclear potential.[15]

What is actually meant is participation in two regional conflicts of small and average intensity, occurring 'almost simultaneously' in terms of time, needing the shipment of troops by transport aviation and sea transport. In connection with the reduction of American troops in Europe and other regions, there appears a likelihood for increased powerful strategic reserve on the territory of the USA.

Special importance is given to the increase in strategic mobility of the armed forces, which is procured by the preservation of the majority of military bases abroad, by the creation of rapid deployment forces comprising up to 300,000 servicemen, and an increase in mobility of the armed forces as a whole through the reorganization of the army and the enhancement of transport capabilities of the air force and navy.

Overall, just like in Kennedy's administration, priority is given to the development of conventional armed forces, with a significant reduction of strategic nuclear forces, which are viewed more as a means of restraint than as the basic means of achieving victory. (In the Reagan administration there were 40,000 targets for nuclear attack and now only about 7,000 targets.)

In the report by secretary of defense L. Aspin, and chairman of the joint chiefs of staff General C. Powell, to accord with the concept of readiness for participation in two wars, estimates were presented for the manning with ground forces (10 divisions of regular troops, 3 marine divisions, 15 brigades and one reserve division) of two attack groupings composed of divisions, and four or five Marine brigades. The Navy was allocated four to five attack carrier groups for each of the prospective wars, while air defense was assigned ten tactical aviation squadrons and up to 100 heavy bombers for each theater of war. Strategic aviation, the aircraft F-117, EE-111, B-2, the AWACS system, J-STAR, naval assault and air transportation facilities can be used as the supreme command thinks appropriate.[16]

In view of the immense complexities of the shipment of troops and large strategic reserves from one theater of war to another and from the home country, not only prior storage of large batches of weapons, military equipment and material stocks in major theaters of war (in Germany, Japan and Southern Korea) is maintained, but new logistic bases in Kuwait, Saudi Arabia, Egypt and other regions have also been organized. The infrastructure, indispensable for this purpose, is expanded for use, if need be, of mobile and peacekeeping forces.

Moreover, according to this strategic concept, the peacekeeping forces are viewed not only from the point of view of participation in the foreign policy and international actions, but also as part of a general plan for participation in local wars and conflicts, where, as in Somalia, they can become a kind of advanced group (echelon), ensuring if necessary, the deployment of the main forces. Forces allocated by the USA for peacekeeping missions total 50,000 servicemen. Groupings can comprise one airborne or air assault division, a light infantry division, a brigade of marines, one or two carrier groups, one or two air wings of mixed structure, special forces units, and combat, rear and technical support.

Taking into account the fact that such forces cannot always extinguish the sparks of military conflict, the USA is counting mainly on multinational forces, as it did in the Persian Gulf region in 1991, and in Somalia. Due to the fact that some NATO countries do not willingly approve involvement of their armies in military conflicts, NATO is looking for new allies, and organizes cooperation with Russia, the Ukraine, Belorussia and other countries, in order to entice them to participate in peacekeeping operations.

Generally, military expenditures in the USA remain huge. According to the *New York Times*, at the present moment the USA spends on defense almost as much as all other countries of the world taken together.[17]

Within NATO, France has the greatest independence in its military policy. Its armed forces are also going through a stage of reduction and reorganization. The main purpose of these measures is provision of readiness for participation in local wars that endanger the national interests of France, and the reduction of state expenditures on defense.

In the military doctrine of France the accent on nuclear deterrence is slightly reduced, and the basic emphasis is put on qualitative perfection of the army and aviation. Alongside measure for enhancement of the efficiency of air, coastal and naval defense, much attention is paid to the reorganization and improvement of territorial defense, and its interaction with other defensive structures and civil administrative bodies. With this aim in view, the system of territorial defense is subdivided into eight regional military commands, and command of the Paris region and is subordinated to three defense zones located in Bordeaux, Lyon and Metz.

The reorganization of the army is planned for completion by 1997. There will be an armored maneuver corps based on the 2nd, 7th and 10th armored divisions. The formation of a joint Franco–German corps is also planned, and for that aim one French armored division will be allocated. 'Rapid Reaction Force' corps will be the major component of the French army.

The program of reorganization of the French armed forces requires the modernization and adoption of new armament systems. As reported in the press, by 1997 the army will cease using modifications of weapons and military equipment taken into service in the 1960s and 1970s. These are:

modernized artillery cannons, 'Leclerc' tanks, fire-support helicopters 'Tiger' and 'Jefo', air defense missile complex 'Mistral', and the satellite communication system 'Arix'.[18]

In recent years, great changes have been taking place in the military art and orientation of the armed forces of China. Military, political and strategic guidelines have acquired more active and resolute character. For instance, China rejected the rules of the concepts of 'active defense' and 'people's war', that provide for a deliberate withdrawal across their own territory, and the destruction of the enemy at the expense of large masses of infantry or guerrilla war tactics. As seen in exercises, China's destruction of an invading group of enemy troops is already planned for the initial stage of military actions, with the aim of restoring the territorial integrity of the country. This is to be followed by the penetration of hostile territory.

In order to overcome enemy aggression, China also considers the possibility of a 'preventive strike' prior to the initiation of war, starting with massive aviation raids against enemy groupings prepared for an offensive. It is stressed that the element of surprise is the major prerequisite for the success of a preventive strike. According to theoretical guidelines and regulations, the basic peculiarities of military art in China are:

- concepts of 'people's war' and the 'triune system of armed forces' (the Chinese army has carried out all operations in close interaction with local troops and national militia);
- orientation of the army to achieve victory over a technically stronger enemy at the expense of the maximum utilization of the human factor, and application of effective ways of combat actions;
- establishing as the main purpose of combat actions the destruction of the enemy;
- use of maneuver combat actions as the basic way to attain the goal of the operation. Thus, positional warfare, playing a minor role, should be combined with maneuver and guerrilla actions.[19]

It is emphasized that the basic content of the military actions of the Chinese army will be strategic offensive operations, with a transition to defense only when directed. At the same time, the possible occurrence of defensive actions is admitted. In case the defensive character of action is chosen, three variants are to be employed to repulse enemy intrusion.

The first is based on the concept of 'people's war in modern conditions' and 'active defense', presupposing the maximum utilization of numerous formations of local troops and national militia and the 'destruction of the enemy in the ocean of people's war'. The second variant provides for the preliminary transition to a strong positional defense along advanced defensive borders to inflict the utmost losses on the enemy, with a transition into the counter-offensive for the restoration of the border.

The third variant, prescribes launching a strike, with the intention of defeating any enemy intrusion.

To repel an enemy offensive, China's guidelines or regulations recommend the formation of three defensive lines: a deterrence line ('strong layer'), an attrition line ('viscous layer') and a line of strategic reserves ('repelling layer').

Regarding the development of arms, China aspires to avoid dependence on other countries, trying not to purchase large batches of weapons or combat equipment, but to get access to advanced western technology and to buy separate modifications of the weapon with the purpose of creating its own military technology. Basic emphasis is made on the formation of the third component of strategic nuclear forces – sea-launched strategic missile – and the perfection of missile forces, and strategic aviation.

China has embraced the creation of operational and tactical nuclear weapons which would give China an advantage over states that eliminate such nuclear weapons.

After the disintegration of the USSR and the Warsaw Treaty Organization, the military, political and geostrategic situation of Russia changed radically. The Russian Federation makes up 76 per cent of the territory, 60 per cent of the population, and 40–50 per cent of the total national product of the USSR. In 1985 the national income of the USSR was 57 per cent of that of the USA, while in the mid-1990s Russia's national income is only 17.2 per cent of that figure. In the geopolitical respect, Russia has been moved from the European center, deep into the Eurasian continent. It has lost access to major ports on the Baltic and Black Seas. Russia's defensive potential has changed considerably; its combat potential for conventional weapons in Europe, limited by the Paris Agreement, will be four to six times smaller in quantity for ground types of arms, and two times smaller for aviation.

Russia aspires to revival as a great power, but is compelled to weigh national interests against existing realities.

Proceeding from the defensive character of its military doctrine, Russia strives, as far as other countries are willing, to ensure the military security of Russia by organizing collective defense with the countries of the CIS. This matter being rather complicated, Russia places main emphasis on the re-creation and reformation of its own armed forces. Military doctrine has assigned special importance to the prevention of wars and military conflicts, rejecting the use of military force as a means to political ends. Russia rejects any territorial claims against other states, though some territorial disputes are yet to be solved, for example the status of Sevastopol.

Russia excludes any possibility of the initiation of war, and in general, the first initiation of any military actions. None of the neighboring states are regarded by Russia as an enemy. Occupying a central place in Eurasian geostrategic space, Russia comprehends its responsibility for the maintenance of stability in this space, and struggles to organize collective defense within the frameworks of the CIS. Russia does not exclude the possibility of deployment of its troops and bases on the territory of other states, as provided by corresponding agreements with these states. The Russian Federation is gradually reducing its armed forces to 1.5 million men, and will actively support international negotiations on the further reduction of nuclear and conventional weapons.

Russia will possess comparatively smaller mobile armed forces, capable of solving two main tasks: 1) the strategic nuclear deterrence of a potential aggressor; 2) readiness for the fulfillment of combat missions in local wars and conflicts, and also to participate in international peacekeeping operations.

In accordance with the set tasks, general purpose forces will probably consist of four basic groupings:

- Western, including troops of Moscow and Leningrad military districts;
- Southern, the basis of which will be constituted by troops of the North-Caucasian military district;
- Central, composed of troops of the Ural, Volga and Siberian military districts, where mobile forces and reserves will be mainly deployed;
- Eastern, including troops of Baikal and Far-Eastern military districts.

From the point of view of strategic basing and the division of military administration, the military districts are so far preserved. But at the same time, the possibility of establishing separate commands for control of these groupings of troops is not excluded.

The role of the mobile forces, deployed in depth and able to react quickly to military conflicts, is increasing. In the opinion of Colonel General A.I. Nikolaev, the role of border-security troops is a major element of the screening echelons. Therefore, they should now be stronger.[20] Under the program of Russian military reform, positive purposes and requirements are put forward that the armed forces be more effective, better oriented within the qualitative parameters, and equipped with the latest precision weapons. But under the present economic situation, this aim will not easily be achieved in the short term. As a whole, Russia and its armed forces are in the process of transformation, but being in a state of social and political uncertainty and internal instability, and having a huge nuclear potential, it is perceived as a source of danger.

Other CIS countries have about the same orientation of military doctrine with due account of national specificity. They depend on Russia for armaments and hardware. Meanwhile, the central Asian and Caucasian republics are establishing military cooperation with Turkey, Iran, China and other neighbors. East European and Baltic republics increasingly gravitate to NATO countries, and are already partially re-orienting their defenses towards the east.

All this creates a completely new arrangement of military and political forces in Europe and Asia.

Thus we may expect that evolution of military doctrines among the leading states of the world in the near future, will develop along the following basic directions:

- non-use of nuclear weapons as a major war-fighting method; the use of nuclear weapons only as a restraint;
- termination of global military confrontation between two coalitions within the states, with a transition from open hostility, considering each other as enemies, to a position of partnership and cooperation; and as a result refusing to prepare for war, whether nuclear or conventional, for the first time since World War II;
- sustaining the readiness of the states and armed forces in maintaining security, and in resolving military problems in one or several local wars and military conflicts, acting whenever possible with joint coalition groupings of troops. At the same time, the armed forces must remain ready to repulse aggression, and to participate in large-scale war, if peace be broken for any reason.

The main emphasis in military structure is now on the qualitative perfection of conventional weapons, the increase of combat power and mobility of forces, and the preparation of armed forces for military operations with the application of conventional means; yet this attitude must still take into account any threat of the use of nuclear weapons, especially from the point of view of a war's outbreak, guaranteed retaliation, and protection of troops against weapons of mass destruction.

Along with the general tendency for a balanced approach to present military–political and strategic problems, there are, as always, extremes and extremism. One extreme is the thought that for the defense of any country it is enough to have a minimum quantity of nuclear weapons as a means of deterrence to guarantee the military security of the state.

In the *Professional army abroad*, in the words of a young French officer referring to Churchill, the 'Defense of our country is secured by nuclear deterrence and will not be paid for by the senseless exchange of human life, glorious and stupid frontal attacks, blood and bodies, piling up before machine-guns.'[21]

Further on, the authors themselves come to the conclusion that 'it is not by chance that the concept of nuclear deterrence underlies the military doctrines of many countries, making direct aggression against that state which has nuclear weapons and convincingly demonstrates a readiness to use them, practically impossible. The effects of nuclear weapons are such, that irrespective to which phase of war it is applied, that phase will be the last. It is obvious, that in this case no reserves will be necessary.'[22]

Such views receive even greater recognition from scholars and politicians. It is sometimes argued that the possession of nuclear weapons makes other services of the armed forces redundant. It should be said that these judgments are based on an enormous misconception. Certainly, nuclear weapons, as long as they exist, will always be a powerful means of constraining aggression.

But it is impossible to ensure a reliable defense with nuclear weapons alone. First, many countries have even now the ability (to be increased in the future) of dealing a surprise strike not only with nuclear, but also with conventional precision weapons, in order to destroy the nuclear bases of other countries, depriving them of the ability to retaliate or carry out nuclear retribution.

Second, armed forces are necessary in at least a minimum quantity, to cover nuclear facilities with air and ground means, and to ensure their combat readiness.

Third, without a minimum of highly trained forces it is impossible to ensure the sovereignty and general defense of a country, let alone the ability to react to various military conflicts.

Therefore, while supporting as a whole the principle of sufficiency of armed forces to provide a defense, it is necessary to say that defense now and in the future can be secured only by a certain balance of armed forces, able to provide sufficient security. Such a sensible approach will also deflect some countries from the illusory temptation to get hold of nuclear weapons at any cost, and to solve their defense problems with it.

Speaking from the point of view of geostrategic stability, it would be most expedient if all states adhered to the principle of sufficiency of defense, when considering the sum total of their armed forces and munitions.

In the theoretical development of criteria of sufficiency of defense, average norms and parameters gained from international practice should certainly be taken into account. It is considered that the size of armed forces should constitute at best no more than 0.7–1 per cent of the population of a country; and total military expenditures 1–3 per cent of the GNP (gross national product). But actual defense requirements can only be determined by the arrangement of military and political forces, the geopolitical situation of the state, the nature of existing military threats and the resulting defensive tasks.

For example, Russia, Belgium, Australia or Syria function under completely different conditions, hence the difference in their defensive tasks; so that in defining a proper structure of armed forces for peace or war it would be silly to rely only on average figures.

Such an approach, which some Russian deputies love to entertain, would be reminiscent of a military medical assistant who took the temperatures of all the sick in a regimental infirmary (with variations from 36–40°C), deduced the average, and on making a diagnosis, took the average temperature as a basis.

But with a serious approach to business, the determination of forces and means necessary for defense requires appropriate estimates of effectiveness, with due account for the specific peculiarities of each state.

How are we to realize nuclear deterrence under these new conditions? What will be the fate of the nuclear weapons, what approaches to their application can be adopted, with regard to everything said up to this point? In the long term, as we know, it is proposed to completely eliminate nuclear as well as chemical, biological and other weapons of mass destruction. However, throughout the next few decades nuclear weapons will still be with us. Even into the next century the USA and Russia will have 3–3,500 nuclear warheads each. France, Britain, China and other countries, which legally or illegally possess nuclear weapons, are not going to abandon them. For despite their monstrosity, nuclear weapons are the most reliable and cheap means to maintain security and to deter aggression, though one cannot but understand that these 'cheap' means could cost mankind dearly. But this is the objective logic of the development of events. Nevertheless, since nuclear weapons still exist, the question remains: what to do with them, how should they be used?

There are various approaches to resolving this conundrum. For example, O.A. Grinevsky, who was engaged for many years in negotiations on the restriction of nuclear arms, supporting the concept of nuclear restraint in general, writes: 'If the Cold War has gone into the past, then the nuclear weapon should not serve as a means of restraining its ghosts.' The 'ghosts' are large nuclear states such as the USA or Russia.

In his opinion, nuclear weapons should be intended for use against 'The states and regimes that profess aggressive nationalism and are ready to use nuclear means in its realization' (so-called 'nuclear adventurers'). He believes that for this purpose it is sufficient to have only a few dozen nuclear weapons, which should be transferred to the direction of the UN security council. The problem of the restriction of 'adventurers' by nuclear restraint actually makes further perfection of nuclear means and nuclear tests needless. The author is convinced that the 'present level is sufficient and even excessive'.[23]

In all probability events will develop in accordance with this variant. But nevertheless, as I see it, in this case the resolution of the nuclear problem is a little bit simplified. O.A. Grinevsky probably proceeds from the variant that future confrontation between great powers will be completely excluded, or that undivided domination of the USA will be established in the world, and there will be no other opposing states. But then (in view of strict international norms of non-proliferation of nuclear weapons) 'nuclear adventurers' will hardly dare to use nuclear blackmail.

As to the great powers, various political and economic contradictions among them arise even at present. In case of further aggravation or the coming to power in some countries of new political forces, other states will be compelled to rely on nuclear weapons and so intend them not just for the deterrence of 'nuclear adventurers'. So in reality, in the foreseeable future the concept of nuclear deterrence should be more multifaceted.

Political and strategic insight and apprehension of the fundamental senselessness of nuclear war should be realized by practical steps on its renunciation by all states, including NATO members, of plans and scenarios of nuclear war; which scenarios lead to the conclusion of agreements, or acceptance of unilateral obligations of the states on renunciation of first use of nuclear weapons, as happened with the USSR and China. In other words, it is necessary to exclude any chances of planning for, and the conscious initiation of nuclear war, for any purpose and under any pretext.

However, the doctrinal guidelines of NATO still stipulate first use of nuclear weapons in response to any aggression, even if it is undertaken with the use of conventional weapons only. Such a concept was formerly justified by Soviet 'superiority' in conventional forces and by the 'Soviet military threat'. But now, when it has all gone, it is difficult to excuse first use of nuclear weapons.

Therefore, when the USA and NATO are criticized by the public and ruling political and military circles of these intentions, they have to admit the vulnerability of their position concerning the application of nuclear weapons. It must have been a great relief to know that Russia has also refused the principle of no first-use of nuclear weapons in the new military doctrine. In the given situation, China will probably be compelled to follow this step. With the publication of the new Russian military doctrine both in Russia and in the West, it was first of all noted that 'Russia, contrary to the USSR, does not exclude first use of nuclear weapons.' Russia 'has rejected the promise of no first-use of nuclear weapons, that was declared under Brezhnev.'[24]

Asked by a correspondent of *Izvestiya*, whether the new doctrine contains an obligation for Russia not to apply a first nuclear weapon, Russian defense minister Army General P.S. Grachev said that none of the military doctrines of the nuclear states has such a definitely expressed position, but that our doctrine includes the obligation not to use it against non-nuclear countries. 'Nothing is said about other countries,' emphasized the minister.[25] Thus, first-use of the nuclear weapon is legitimized as the only possible way of its

The concentration of tanks and APCs for the parade after the 1980 military exercise in the Belorussian Military District. Under modern conditions, such a concentration of armor would be an inviting target for advanced precision fire and strike complexes.

General-Colonel Gareev with Chief of the Soviet General Staff, Marshal Nikolai Ogarkov, during the 1982 military exercise of the North Caucasian Military District. Marshal Ogarkov supported the development of Gareev's concept of the operational maneuver group in the late 1970s and early 1980s.

General-Colonel Gareev with Chief of the Soviet General Staff, Marshal Sergei Akhromeev, greeting the Chief of the Turkish General Staff, General N. Urug during an official visit to Moscow in November 1985. To Akhromeev fell the difficult task of negotiating the military disengagement in Central Europe which culminated in the treaties on Intermediate Range Nuclear Forces and Conventional Forces in Europe.

General-Colonel Gareev with Marshal Viktor Georgievich Kulikov during military exercises in the Belorussian Military District in 1986 just as perestroyka was beginning.

General-Colonel Gareev among the Soviet delegation greeting US Secretary of Defense Frank Carlucci during an official visit to a higher military–educational institution in Moscow in June 1988, just after official contacts between American and Soviet military officials were restored in the wake of the withdrawal of Soviet forces from Afghanistan.

General of the Army Makhmut Akhmetovich Gareev in his office at the General Staff of the Russian Armed Forces in the early 1990s.

Strategic Operation (One Option)

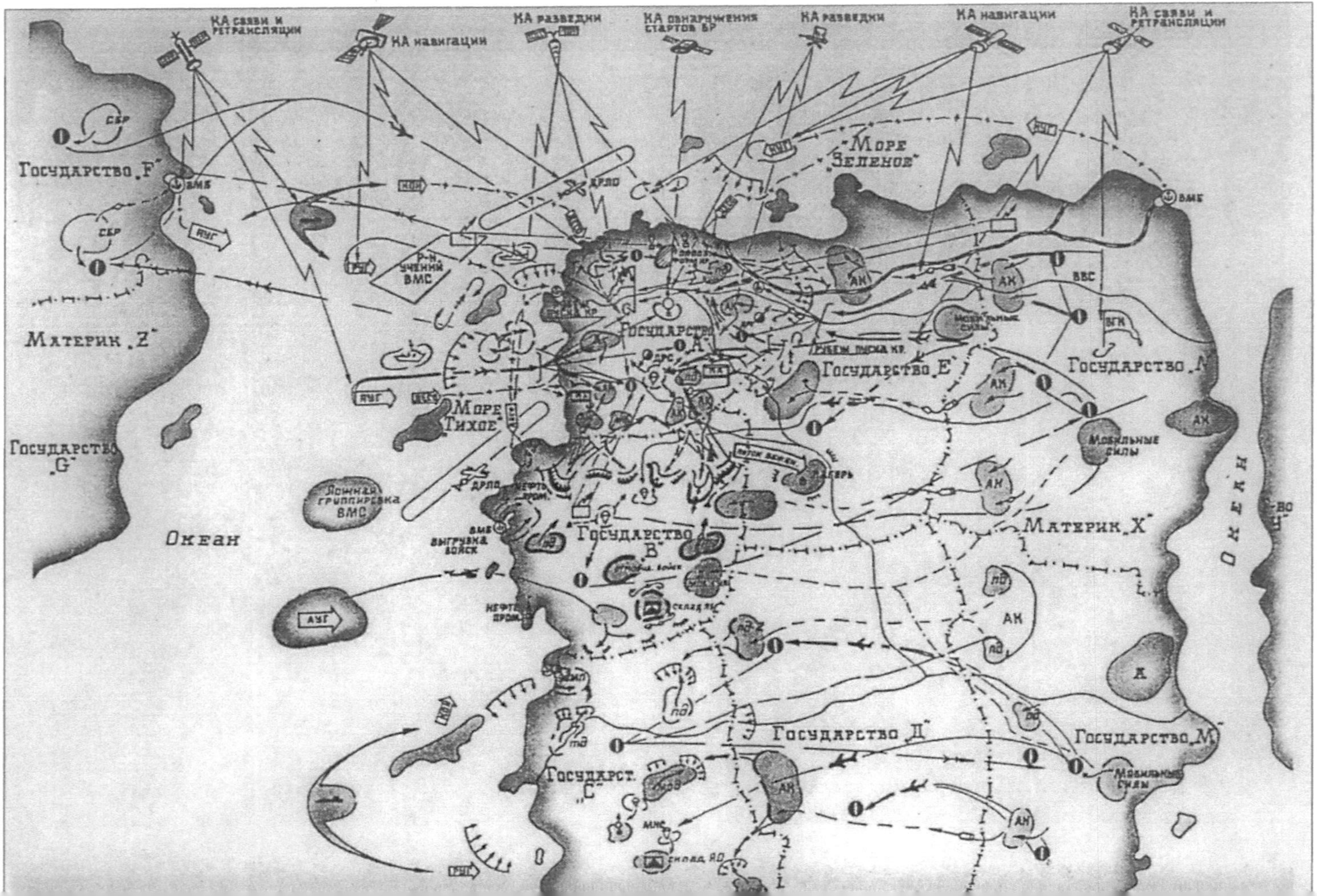

Государство «F» – State "F"
СБР– strategic ballistic missiles
Материк «Z»– Continent "Z"
Государство «G» – State "G"
АУГ – carrier task force
ВМБ – naval base
КА связи и ретрансляции – communications satellite
КА навигации – navigation satellite
КА разведки – reconnaisance satellite
КА обнаружение стартов БР – ballistic missile launch detection satellite
РУГ – cruise missile attack group
КОН – convoy
Р-Н УЧЕНИЙ ВМС – naval exercise area
ДЕСО – amphibious assault group
ДЕЛО – long-range radar
АУГ – carrier task force
МОРЕ «ЗЕЛЁНОЕ» – Ocean "Green"
ВВС – aviation forces
АК – army corps
ОППОЗ. ФОРМИР. – opposition formation
РУБЕЖ ПУСКА КР – cruise missile launch perimeter
ГОСУДАРСТАВО «А» – State "A"
ПД – infantry division
МОБИЛЬНЫЕ СИЛЫ – mobile forces
ВГК – Supreme HQ
ГОСУДАРСТВО «N» – State "N"
ГОСУДАРСТВО «Е» – State "E"
ПА – field army
ДРС – radio communications center
МК – mechanized corps
ЛАГЕРЬ – camp
ПОТОК БЕЖЕН – flow of refugees
НЕФТЬ. ПРОМ. – oil industry
ДРЛО – long-range radar
ЛОЖНАЯ ГРУППИРОВКА ВМС – false naval group
ВЫГРУЗКА ВОЙСК – troop disembarkation
ОТМОБИЛ. ВОЙСК – mobilizing troops
ТД – tank division
Д. МОБ. СИЛ – concentration of mobile forces
МАТЕРИК «Х» – Continent "X"
ПД-2 – two infantry divisions, concentrated
СКЛАД ЯБ – nuclear storage facility
НЕФТ. ПРОМ. – oil industry
ЭДМП – marine division element (naval infantry)
А – army
ГОСУДАРСТВО «М» – State "M"
МОБИЛЬНЫЕ СИЛЫ – mobile forces
ГОСУДАРСТВО «Д» – State "D"
МОВ – mobile forces
МНС – multi-national forces
СКЛАД ЯО – nuclear storage facility
ОКЕАН – ocean
Государство «У» – State "U"

Forces and Means

Unified space forces of States "F", "M" and "N"
Unified air forces of States "F", "M" and "N"
Unifed naval forces of States "F" and "N"
Mobile forces of States "M" and "N"
Rapid deployment forces of State "F"
Unified ground forces of States "B", "F", "M" and "N"
army corps 5–7
infantry divisions (mechanized infantry divisions) up to 9

Operational Situation

1. State "A" carries out aggression against State "B" with the objective of seizing oil fields. State "C" supports "A" and threatens the use of nuclear weapons. States "E" and "D" assume a wait and see position but secretly support states "A" and "C".
2. Upon the decision of the United Nations, States "F", "M" and "N" conduct an operation to repel aggression and liquidate the conflict.

Possible Stages of the Operation

First Stage

1. The initiation of political actions aimed at preventing war:
   - negotiations, warnings, political pressure and a search for compromise
   - supporting forces, acting against the ruling regimes in the states in conflict.
2. The execution of economic sactions and naval, air and ground blockade of the lines of communication of the warring sides.
3. Demonstrations of force. The conduct of psychological and special operations.
4. The execution of joint and mutually-agreed upon measures concerning disinformation regarding the nature of the anticipated actions of the multi-national forces. The planning and preparation of the armed forces for their employment. Covert deployment of the armed forces for the anticipated actions.

Second Stage

1. The beginning of combat actions against State "A". The conduct of an air operation of 20–25 days duration employing the air force, naval forces, special operations forces, mobile forces, electronic warfare and intelligence units and aero-space means of support with the objective of carrying out fire destruction and electronic suppression of enemy command posts, and air, naval, and group groupings of forces, the disorganization of state and military command and control and the breaking of the aggressor's will to resist.
2. Simultaneous shifting, regrouping and movement of mobile forces, tactical aviation and basic groupings of the ground forces into the regions of anticipated actions.
3. Presentation to the aggressor of an ultimatum concerning his capitulation or other terms for the cessation of military actions.
4. Warning to State "C" about the possibility of nuclear attack in case of its employment of weapons of mass destruction.

Third Stage

1. Execution of combined air–land operation against the aggressor. Introduction into the engagement of mobile forces and the basic groupings of the ground forces with the execution of fire strikes on strategic and operational targets through the depth of the territory seized by the aggressor. Simultaneous envelopment of the groupings of his armed forces from the air and sea, seizure of bases and regions used by his naval and air forces, of his energy infrastructure and military–economic potential, and the command and control facilities of the state and military.
2. The blockade of the groupings of armed forces of State "C" and the seizure of its strategically important facilities and centers for the production of nuclear weapons.
3. Protection of refugees, material and medical support for them in areas of their mass concentration.
4. The institution of measures for the reestablishment of state authority of the warring sides and the prevention of the repetition of the armed conflict.
5. Withdrawal of the armed forces of the intervening states.

General-Colonel Gareev as Senior Soviet military advisor to the government of Afghanistan after the withdrawal of Soviet forces. In this photograph, taken in January 1990, General Gareev is meeting with a mujahadeen faction that has agreed to cease hostilities against the government. Local wars, like that in Afghanistan, appear to be one of the key trends affecting the contours of future armed conflict into the next century.

utilization. Some western military experts explain this step of Russia's by the fact that earlier proclaimed principles of the non-use of nuclear weapons had a purely propagandizing character, and no practical importance.

Besides, as some experts believe, in view of the diminishing general military power of Russia, it can now basically only count on the nuclear weapon in safeguarding its security. The West has reacted positively to this, probably proceeding from the assumption that the threat of the Russian nuclear weapon is not directed against it although the application of nuclear weapons is a threat for everyone. Presumably, there must be some other reasons.

But if we address this problem from the point of view of the interests of global security, in the long term it is obviously necessary to take into account the following circumstances. If only the military and strategic aspect is taken into account, the determination and readiness for first-use of nuclear weapons in response to any aggression makes nuclear deterrence most effective, both from the point of view of warning a potential aggressor, as well as in the reliability of the use of nuclear weapons, for any hope of retaliation is obviously forfeited from the military point of view.

However, from the point of view of politics and the interests of global strategic security, the principle of first-use of nuclear weapons has essential negative aspects.

First of all, in any aggravation of the international situation, the rush not to be last and the rush to use nuclear weapons first will lead to dangerous competition, complicating the situation and provoking a preventive nuclear strike. Under such conditions it will be difficult to determine who used the nuclear weapon first and for what reason. It creates the premise for various sorts of risk taking, and for pulling a screen over aggressive actions. Preventive actions can be undertaken and nuclear weapons used in 'response' to invalidate the aggressive 'intentions' of one or another state. Suffice it to recall, even the completely obvious aggression of Hitler against the USSR in 1941 is still doubted by some circles. The initiator of aggression can always find a pretext for, and conceive justification for, its actions.

Speaking realistically, who might be the target of Russia's first-use of the nuclear weapon? In case of aggression by states surrounding the Russian Federation and the CIS, it will probably be able to solve defensive problems without using nuclear weapons. If we assume hypothetically that it will be necessary to use these weapons against the USA and other NATO countries, it can only result in global catastrophe. Then only China is left. But China has undertaken the obligation not to use the nuclear weapon first. Besides, massive employment of nuclear weapons on the territory of Eurasia, where Russia is situated, threatens larger disasters than its application against the West.

Some military experts assume that the possibility of first use of nuclear weapons is proclaimed by Russia in order to intimidate some obstinate neighboring states. But such an approach can result only in mutual suspicion and hostility, complicating the maintenance of military security.

An intention to assume the strategic initiative, and other purely military principles, are not always applicable for the nuclear weapon. In this field, political approaches to the prevention of the use of nuclear weapons are especially important.

In this age of nuclear missile weapons, an absence of any limiting agreements and obligations creates the premise for a much greater arbitrariness and manipulation with regard to nuclear weapons. Thus, any conclusion of international agreements on the prohibition of first-use of nuclear weapons and the appropriate obligations of nuclear states (even in the absence of the complete guarantee of their observance) would serve as a certain legal and moral barrier to constrain to some degree the use of nuclear weapons. This is especially important as massive use of nuclear weapons under any circumstances is senseless for both sides, and because the main purpose of nuclear weapons nowadays is not their usage, but their political means for deterrence of aggression.

The guidelines for first-use of nuclear weapons contradict even the defensive characters of military doctrines.

On the one hand, the first part of Russia's military doctrine states, and is confirmed by former minister of foreign affairs A. Kozyrev, that 'military force can be applied only with the aim of self-defense, for rebuff of aggression ... we are not going for any kind of first-use, and regard nuclear weapons as the most extreme means'.[26] On the other hand, a possibility of first-use of nuclear weapons is not excluded. It turns out that if infantry and tanks cross the border of another state it is an aggression, but if the same state is attacked with nuclear weapons it is not aggression, but something else.

One can hardly agree with the references to the fact that obligations of no first-use of nuclear weapons have only a propagandistic importance. With such a stance it would be possible to announce as purely declarative even the guidelines of the military doctrine on non-use of the nuclear weapon against non-nuclear states, including any other agreements and obligations. History knows of many cases when contracts and agreements were treacherously broken, but it also reveals examples of their observance, including international regulations concerning non-use of chemical weapons during World War II.

If international control for the planning and application of nuclear weapons would be established, and agreements directed at realization of this control concluded, it would add to guarantees for restraining the initiation of nuclear war.

The fact is, in a case of the elimination of ICBM (inter-continental ballistic missiles) with MIRV (multiple independently targeted re-entry vehicles) Russia loses the possibility for delivering a preemptive counter-force strike.

The interests of the stabilization of the global strategic situation and the nature of mutual relations in the strategic sphere require the possibility for the repudiation not only of a first preemptive strike, but also a retaliatory attack.

From the military and strategic point of view, an alternative to preemptive nuclear strike, and one of the ways to increase the effectiveness of nuclear deterrence, can be the provision of a guarantee for reliable nuclear retaliation. This point should be achieved at negotiations on strategic offensive forces, so that no party to the agreement loses the possibility for delivering a retaliatory nuclear strike. For this purpose it is also very important to observe the restrictions set on the deployment of antimissile defense systems.

In view of the above, there is some reason to believe that principles of the use of nuclear weapons will in the future be specified and improved, so as not to facilitate, but on the contrary to limit, and finally not to permit, the use of nuclear weapons under any circumstances. All states should also reject concepts of 'limited nuclear war' and 'limited nuclear strikes', because they can provoke nuclear war.

A readiness to retaliate with nuclear weapons should be sustained only in the case of use of those weapons by the enemy. In this connection, the limited planning for the use of nuclear weapons will be preserved, and armed forces will be trained to retaliate and defend against weapons of mass destruction. It is sometimes argued that any planning and preparation for the use of nuclear weapons should be banned. But this is unreal, for then any sense of the existence and use of this weapon is lost. It is also the fact that without some preparation for the use of nuclear weapons it is impossible to ensure even the threat of retaliation.

Whatever the catastrophic consequences of nuclear war might be, none of the armies are likely to remain idle and surrender at once when facing this danger, but will probably aspire with any remaining forces to continue missions in order to inflict considerable damage on the enemy. Only the combination of preparation and determination against such actions can serve nuclear weapons as a means of reliable deterrence against potential aggressors.

To prevent one side from winning, not only the attackers, but also the opponents need to be effective. Otherwise it is impossible to deflect an aggressor from the temptation to gain a victory, even an illusive one.

Renunciation of the concept of global nuclear war, planning only for constraining nuclear strikes, shortening combat duty, the introduction of additional measures of trust and mutual control in control stations and in outer space, will all assist the transition into the preliminary stage of the post-nuclear epoch.

The key question still is: which objects should be the prime targets for nuclear strikes? It is considered that nuclear weapons can perform restrictive roles only when aimed at large industrial cities. Delivering nuclear strikes upon strategic nuclear targets is meaningful only as a preemptive counterforce strike, because it is almost useless to destroy nuclear launchers after they have delivered their strike. L.S. Semeiko, considering this point with reference to Russia, writes: ‘On the one hand, it is envisaged that Russian nuclear weapons will serve exclusively as a means of deterring large-scale war. On the other hand, it is supposed (in view of the official declarations), that Russia will not strike against cities of the USA and western Europe, whereas just the threat of such strikes is the main factor of deterrence ... Avoiding strikes on small-size and fortified strategic objects assumes in itself a shifting emphasis to the so-called countervalue employment of strategic nuclear forces, i.e. readiness to destroy the greatest values – cities and major industrial centers.’[27]

Such a barbarously severe approach to the use of nuclear weapons could somehow be understood when two states of about equal nuclear force are confronting each other and are determined upon global nuclear war. Now, when the global confrontation of two superpowers is, at last, over, it is impossible to regard such an approach as the only possible solution. In this connection it is imperative to study other variants of actions, for example an interdiction of nuclear blackmail, or nuclear strikes with terrorist purposes, that could occur in the future. In using the nuclear weapon against an aggressor state possessing a small amount of nuclear weapons, and even if this state strikes first, it is more expedient to deliver retaliatory strikes against nuclear installations, and not against industrial centers, where a largely peaceful population would be destroyed. Besides, if we admit the theoretical possibility of the limited application of nuclear weapons in the military confrontation of two major nuclear states, then in this case it might be appropriate to deliver retaliatory strikes not against cities and civilian populations, but against enemy military-strategic targets. Other variants of such actions could also occur.

Even if war commences using conventional weapons, strikes against targets of strategic nuclear forces, nuclear power stations and other especially important objects, can provoke the use of nuclear weapons. Irrespective of whether any agreements regulating the actions of the parties in regard to nuclear targets will be concluded or not, the opposing sides will always be compelled to take into account the danger of engendering nuclear war.

In the event of the future elimination of nuclear weapons, modern technology easily permits the rapid re-creation of nuclear weapons, so that suspicions will be constantly arising that the opposing party might restore its nuclear potential. The shadow of nuclear weapons will always hang over mankind, compelling the equation of the political and strategic expectations of states with the possibility of the materialization of nuclear war.

Transitions to new concepts of nuclear deterrence cannot be limited to political declarations only. Their practical realization requires changing the system of nuclear planning, and orientation of the armed forces. But so far little has changed in this respect among the nuclear states. For example, in the USA the planning of the potential use of nuclear weapons is still officially determined by the 'National Security Directive No. 13', signed as far back as 1981 by R. Reagan, and directed at the strategy of preparation for and conduct of global nuclear-missile war. This very directive was taken as the basis for the 'Strategic Integrated Operational Plan' Siop-5p, which contained various directions for the destruction of major military, economic and other state targets in the USSR and countries of the Warsaw Treaty Organization, and included 40,000 potential targets for nuclear strikes. As the American military expert W. Arkin noticed: 'All our actions only immortalize the long-standing race in nuclear arms.'[28]

Former systems of nuclear planning and the preparation of armed forces in general, are still used today in Russia, China and other nuclear states. Combat regulations and training programs for troops and military academies all describe measures for the preparation and conduct of combat operations under conditions of the use of nuclear weapons.

With regard to this, a rejection of the strategy of global nuclear war and the realization of new concepts of nuclear deterrence, require the realization of a whole complex of practical measures. To my mind they could basically be as follows:

- further reduction up to a necessary minimum of strategic nuclear forces, and the complete elimination of tactical nuclear weapons including air capabilities;
- official declarations by all nuclear states on the renunciation of the strategy of nuclear war and first-use of nuclear forces, and a transition to the concept of nuclear deterrence, preparing forces only for retaliation;
- conclusions of international agreements on the above-listed matters, with participation by all nuclear states;
- strengthening and tightening international control over the non-toleration of the further proliferation of nuclear weapons, plan for their use, maintenance of nuclear security, and the organization of combat regimen of strategic nuclear forces. It might be appropriate

to organize special bodies within the military and staff committee of the UN Security Council, designed to control the observance of norms and restrictions set by international agreements on nuclear planning. The problem of rapid interdiction of attempts to use nuclear weapons for terrorist and other purposes should be given special attention;

- reorganization by all states of preparation of systems of the armed forces, reassessment of combat regulations and training programs for ground, air, and naval forces and military educational institutions, with guidance away from nuclear-missile war, and towards fulfillment of combat missions with the employment of conventional weapons.

Other purely military measures are also necessary in order to change the orientation of nuclear planning, military structuring, and the combat preparation of national armed forces. Specifically, under conditions of significant nuclear arms reduction and the rejection of preemptive nuclear strikes, strategic stability and the security of the states can be provided by the deepening of measures of trust and mutual control.

Essentially, these can be as follows:

- coordinated efforts on the perfection of early warning facilities for the detection of missile, air and space strikes, exchange of data about any danger of nuclear aggression (joint launch and use of space satellites, sea and ground early warning stations), and for the future the creation of a global surveillance system;
- information exchange in the field of nuclear planning, with the creation of missions at general headquarters and control points of strategic nuclear forces;
- acceptance of a number of coordinated measures on the reduction of the possibility of the surprise employment of nuclear weapons by forces in neutral (safe) regions;
- removal of nuclear systems from combat readiness, subject to their reduction in accordance with agreements; removal of all nuclear charges from strategic aircraft, and their storage in mutually designated regions; allowing for combat duty only some single-stage missiles and SSBNs.

But the main thing, of course, is strengthening measures of trust, like those we see in the example of the mutual relations of the nuclear states of Great Britain and France, which do not even presuppose mutual nuclear deterrence.

All the above listed measures for the reduction and overcoming of nuclear danger, are intended for a situation where the USA, Russia, China and other nuclear states, will in due course conclude agreements, and accept obligations on their fulfillment, and will also establish effective international control for the maintenance of nuclear security.

In case this does not happen and some states continue to rely on nuclear weapons, and moreover, to increase them, the preparation for nuclear war in one form or another will proceed. However, sooner or later common sense will probably prevail and for the sake of its salvation, the international community will be compelled to find ways for the provision of nuclear security.

Therefore, despite discrepancies permeating tendencies in the development of the nuclear problem, in the light of what has already been said, there is every reason to believe that nuclear strategies will die out and that nuclear war will become more and more improbable. In this connection, military art will in the future be guided by the possibility of the occurrence of conventional wars. But nuclear weapons will continue to render their decisive influence on strategic stability in the world, and on the political and strategic missions, the course and outcome of conventional wars, even when this weapon is not applied.

The possession of nuclear weapons by antagonistic sides, both during a pre-war confrontation and in the course of a conventional war, should deter both sides from risking excessive requirements, or to expand their military actions, or to pursue conquest of all, which would provoke the use of the nuclear weapon.

2 METHODS OF CONDUCTING OPERATIONS AND COMBAT ACTIONS IN CONVENTIONAL WAR

Given the huge destructive power of modern and prospective armaments, even conventional war (without application of nuclear weapons) will be more destructive than former wars. Especially dangerous will be the destruction of atomic power stations, chemical plants, and similar objects. Sometimes the opinion is expressed, that such consequences could be as grievous as that of nuclear war. In reality, if one or more nuclear reactors (there are about 150 of them in Europe) is destroyed by conventional ammunition, the radiation effect will be like that of Chernobyl. Therefore, the conclusion is drawn that in Europe it is impossible to employ either nuclear or conventional weapons. Without underestimating the consequences of war with the employment of the latest conventional weapons, at the same time, in forecasting the nature of future armed struggle, one should not entertain unreasonable exaggerations.

First, conventional weapons are in fact being used; for example in the Caucasus and in Yugoslavia. Second, during the Second World War Germany had no lesser number of chemical factories, hydro-units and dams than now. The majority of them were destroyed by Allied aviation. The consequences were heavy, but not catastrophic. During the Gulf War, the coalition forces' aviation delivered powerful strikes on the nuclear and chemical targets of Iraq. The destruction was enormous, but certainly not as great as if nuclear strikes had been employed.

For a correct prognosis of the nature of future war, not propaganda, but the realistic estimation of possible consequences is necessary.

The Classification of Wars and the Methods of Strategic Employment of Armed Forces

What kinds of war might occur in the future? According to the existing USA classification, there are four versions of wars: general nuclear, general conventional, nuclear and conventional war on the theater of military actions. Concepts of military conflicts of various intensities are also developed. War using nuclear weapons and also war with conventional means are attributed to high-intensity conflicts; while wars conducted for limited purposes are considered to be average-intensity conflicts. All other probable military collisions are referred to as low-intensity conflicts.

Such an approach to the classification of war proceeds from a superstate position. For example, it is not taken into account that if war in Vietnam for the USA was 'a conflict of average intensity', for Vietnam it was a war crucial to the fate of the country.

In Russia, war is differentiated by:

- scale: world war, theater war, local wars, military conflicts;
- weapons used in armed struggle: each type of war (except the last) can be either nuclear or conventional only;
- political ends and international legitimacy: just, unjust, aggressive or defensive (directed at the protection of the Fatherland).

Authors of other classifications suggest a sharp reduction in the number of wars, because preparation for each kind of war requires huge expenditures. They suggest, for example, a withdrawal from 'outdated conventional world war'. Others urge the non-recognition of any local war.

But the authors of these classifications seem to think that the ideological rejection of war might effect its disappearance. But in reality we have not managed to prevent local wars, and the probability of the initiation of large-scale wars, though decreasing, cannot be completely excluded in the long run. Despite the undesirability of war for this or that country, war can simply be imposed as a result of aggression, and then there will be nothing left to do but defend or surrender.

Therefore, the leading states aspire to structure and prepare their armed forces so as to maintain readiness for the resolution of problems in large-scale war, because today it is practically impossible to organize an army for a local war, and then tomorrow suddenly to use it in other capacities.

So it is completely obvious that, under modern conditions and in the near future, the occurrence of local wars and conflicts is most probable. This means that a prime readiness of armies and navies for participation in local wars is required. In this connection, the role of mobile forces is growing. But these forces alone are insufficient. Mobile forces (airborne troops for example) can conduct independent combat actions only for a short period, and then they require support. The USA has strong forward base groupings, and maintains strategic aviation and naval forces, but even these should be strengthened.

For Russia, it should be noted that its geopolitical situation, enormous spaces, underdeveloped system of communications, lack of air, sea and ground transportation facilities for shipment of mobile forces to conflict regions, all extend the time factor for combat operations. Many cities and major economic and military targets are located close to its borders, and require reliable protection. An example from the 12th border post of Russian troops on the Tadjik–Afghan border in August 1993, which for 16 hours independently conducted combat actions without any support, has shown that mobile forces cannot always arrive in the necessary regions as quickly as conditions require. Taking this into account, and in order to be ready for any sort of incidents, it is expedient to deploy in the major directions not only mobile forces, but also the necessary minimum of troops, aviation, and the forces and means of air defense. Successful accomplishment of combat missions in unexpectedly occurring situations depends not so much on structure and arrangement of troops, as on good control, the interaction of border units, advanced units, and troops arriving from the interior of the country, and a high degree of readiness of transportation facilities.

Even in local wars it is not always possible to make do with mobile forces only. Experience shows that for the achievement of combat tasks even in local war, sometimes rather large forces are necessary. For example, by the end of the war in Korea [in the early 1950s], the number of troops involved on both sides reached 2.5 million servicemen, nine times the figure at the start of war. The attempt to resolve problems by rather small forces, as for example in the Soviet–Finnish war of 1939–40, or the introduction of Soviet troops into Afghanistan, led to an expansion of the conflict, and the prolongation of military actions, which in turn increased losses.

If we divert our attention away from political aspects, then the purely military conclusion to be drawn from the war in Afghanistan, or other local wars, is that states or coalition forces, acting under the aegis of the UN or independently, should, for the purpose of the interdiction of aggressive conflicts, act with resolution and concentrate a sufficient number of effective forces capable of accomplishing their missions quickly, and under the right terms. Operations of the coalition forces in the Gulf War of 1991 are a good example of such interaction.

In view of the above statement, the most urgent problems facing military art are research and development in the methods of armed struggle, in local and conventional large-scale wars, and under the constant threat of the application of nuclear weapons.

Up until now, armed struggle was basically considered as a system of strategic operations. But in connection with serious changes in the character of military threats and armed struggle, it already cannot be reduced to strategic operations only. Therefore it is essential to specify the probable forms of the combat application of armed forces.

Basically, they could consist of:

- strategic deployment (mobilization, operational deployment and regrouping of reserves); partial for local wars and conflicts; complete for large-scale war;
- application of armed forces in local wars and in military conflicts. The scale of armed struggle should be met with combat actions and operations of both mobile forces and covering forces, in coordination with the airforce, air defense troops and navy, in continental strategic regions and directions; in combat actions and operations in repulse of an enemy air assault; in airforce operations; in the combat actions and operations of naval forces;
- employment of armed forces in large-scale war as the conduct of operations of strategic nuclear forces; strategic operations to repel air and outer space assaults from the opponent; strategic operations on the continental theater of war; strategic air operations and campaigns; strategic operations on sea and oceanic theaters of war.

In the future, strategic deterrence will probably be viewed as a complex of measures for the prevention of nuclear aggression and planning of nuclear strikes. Strategic operation of repelling air and outer space assault of the opponent will most likely merge with combined-arms, air and naval operations, conducted for the task of destroying enemy air, outer space, and missile groupings, and beating off their assaults.

Thus, in large-scale war of the future, the following kinds of strategic actions are feasible: strategic deterrence by strategic nuclear (offensive) forces; air to air defense operations for the destruction of air and outer space systems of the opponent, and a repulse of their assault with participation of all services; operations of mobile forces and major combined-arms formations on the continental theater of war; and operations at sea and in oceanic theaters of war. Probably, only the naval forces of the USA and NATO will be capable of conducting strategic operations in the theater of naval operations. For the Russian Navy, the carrying out of full strategic operations in a maritime theater will be excessively difficult. This creates completely new conditions for the actions of US and NATO armed forces in the theater of naval operations.

A major point is that reduction of arms has so far least touched the naval forces. The disintegration of the USSR and the weakening of the Russian Navy gave a large advantage to the USA and NATO countries as regards naval forces, creating an essential imbalance in naval strength. An unwillingness to include in negotiations the issues on the reduction of naval forces (while the ground forces are sharply reduced) is usually justified by arguments that naval forces, contrary to ground forces, are not able to seize territory, and consequently, they have a seemingly less aggressive nature. But one can hardly agree with this, if we recollect the fact, that it was the American naval forces that played a major role in conquering numerous islands in the Pacific zone, and the occupation of Japan during World War II, or the role of the British naval forces in conquering the Falkland (Malvinas) Islands. Besides, under modern conditions, sea and ocean communications are less vulnerable than ground ones, and consequently, fast transport by sea of large forces can break correlation of forces as set by the Paris Agreements, and create a threat in some regions. In this connection, the issue of imbalance of sea power in the multipolar world, upsetting strategic stability, still awaits to be solved.

On the Strategy of Direct and Indirect Actions

Speaking of the changes in the character of armed struggle, it is necessary to look anew at a ratio of so-called direct and indirect actions. History, including military history, sometimes repeats itself in a qualitatively new form.

The Chinese military thinker Sun-Tsu asserted that he who knows how to conduct war, conquers an enemy army without a fight; takes enemy castles without a siege; and breaks enemy states without keeping his troops out for a long time.

In fact, during the dynasty wars throughout the fifteenth to eighteenth centuries in Western Europe, indirect actions became ends in themselves. These wars resembled training maneuvers more than military actions, for the contenders tried to avoid battles which could result in the loss of expensive hired armies.

But nevertheless, for centuries commanders recognized that direct actions should only be resorted to when all possibilities for indirect action are exhausted. It is well known, for example, how skillfully M.I. Kutuzov combined diplomatic and military means with various combat actions in battles under Ruschuk and Slobodzeja, that ended with the conclusion of the Bucharest Peace Treaty with Turkey in 1812, which favored Russia, and which occurred before the invasion of Napoleon into Russia.

One vivid example of a successful indirect action was the sending of two Russian squadrons to the shores of America in 1863. The objective was not only to assist northern states during the American Civil War, but mainly to protect Russian interests in Europe by preventing intervention of the west-European states into Russia's internal affairs in connection with the Warsaw revolt of January 1863. As a result of the skillful actions of squadron commander Rear-Admiral A.A. Popov, the entry of the squadron into the San Francisco region posed a threat to British trade communications, and became a turning point in favor of the north. Great Britain and France were compelled to defer any attempts of armed intervention in Russian affairs. Thus, by creating a threat to the rear of the opponent at a vulnerable point, the threat to the western borders of the Russian empire was removed without even a fight. History contains other similar examples.

The situation has now changed, due to the appearance of massive armies, formed on the basis of conscription and general military draft.

Generalizing over the experience of the Napoleonic wars, Clausewitz and Jomini came to the conclusion that the outcome of war is not determined by fruitless maneuvering and capture of territory, but that destruction of enemy armed forces certainly serves as a strategic purpose of national war. In his famous book *On War* Clausewitz proved that war treads only one path, that of combat and battle, and that only large battles of a general character yield great results.

Clausewitz wrote: '[the] destruction of hostile armed forces underlies all military actions ... Bloody settlement of a crisis, aspiration for the destruction of hostile armed forces – this is the first-born son of war.'[29]

These rules, correct to a certain extent at the time, became dominant throughout the nineteenth and twentieth centuries. The followers of Clausewitz's teaching – Moltke, Schlieffen, Ludendorff, Foche, and later Soviet military theorists have transformed his ideas about the key importance of combat and battle into an end in itself, shifting to the background all other methods of achieving victory. The formal use of his principle of the application of force without any restriction or account of what it will cost (a point, which, by the way, he never made absolute), has gradually led to the appearance of the theory of 'absolute' or 'total war', which became a basis of the strategic concepts of the First and Second World Wars, and even in postwar military art.

In the 1920s the outstanding Russian military theorist A. Svechin spoke against an absolutization of the theory of total war, and asserted the need for a combination of various forms of war – war of annihilation and war of attrition, the latter including not only defensive actions in the broad military–political understanding, but also elements of 'indirect operations'. At the beginning of the 1930s he wrote that for the USSR only a war of attrition with limited objectives is expedient, and that the time for the proletarian war of annihilation had not yet come. Svechin argued that

Lack of attention to defense observable in the Red Army is based on a misunderstanding of the dialectic connection between them [defense and offense]; he who cannot defend, will not be able to attack; one should with the aid of a strong defense be able to create the preconditions for attack, the right to attack must be earned.'[30]

In those days, these judgments of Professor Svechin were rejected with sharp criticism, but the year 1941 confirmed his warning.

After World War II B.H. Liddell Hart, author of *Strategy of the Indirect Approach*, was one of the first military theorists to speak out against the theory of total war. In particular, he wrote that adherence to the strategy of total war is 'denial of the art of state control and reasonable strategy, which tries to serve the purposes of politics'. The main idea in his book is the necessity to repudiate the principles of total war, and to revive in politics and the military, the art of the strategy of indirect actions. He emphasized that 'strategy will be most perfect, if it ensures achievement of the purpose without serious combat actions'.[31] But he viewed the strategy of indirect approach in an excessively broad sense, meaning the evasion of decisive battles; waiting for blunders of the opponents; the refusal of frontal actions; and the drawing of sudden strikes against enemy's weak points from unexpected directions; recognition of the key importance of maneuver in the theater of military actions; political measures weakening the enemy rear; the use of cunning new methods of fighting, and still others. In essence, all the major principles of military art and their skillful application were pertinent to 'indirect approach'.

Liddell Hart tried to prove that all the great commanders of the past (from Alexander the Great to Napoleon) and even Hitler in his seizure of Western Europe in 1939–1940, and also the USA and Great Britain in World War II, all largely conducted the strategy of indirect actions. It is hardly possible to completely agree with this though. Of course, even in the past a frontal collision was less preferable than a detour of flanks or an engagement of the enemy in the rear, breaching his communications, but this does not make strategy 'indirect'. As a rule, decisive victory in war has usually been gained precisely by victory in large fierce battles and the destruction of the opponent, followed by the skillful political binding of military successes. Under the system of the anti-Hitler coalition during World War II, Western Allies managed to limited themselves to indirect actions for a long time, but only because major forces of the common enemy were involved in intensive battles on the Soviet–German front.

Overall, despite some exaggerations, and broadness in the interpretation of the strategy of indirect approach, Liddell Hart's book forced the critical assessment of total war experiences for the twentieth century. And it displayed a broader look at the whole diverse arsenal of military art, gained from military history.

In view of the changed conditions of the occurrence and conduct of wars at present, the flexibility of military–political and strategic actions, and the use of a greater diversity of direct and indirect actions, become especially urgent and promising. It is probable that the relative weight of indirect actions will increase. This will be promoted by nuclear deterrence, and the tendency for the maximum preservation of professional armies, and also by the refusal of direct support of conflicts by the great states. In the Cold War period the parties participating in military conflicts (for instance Korea, Vietnam, the Middle East or Afghanistan) usually received the support of one of the superstates, and consequently, the conflict lasted for many years. Under the new conditions, in cases where the majority of the great states act from a joint position, the measures on localization of the conflicts can be more effective.

Indirect actions can be expressed first of all in political efforts on the prevention of wars and military conflicts. As large-scale wars can initiate as a result of the gradual involvement of states in military conflicts and their expansion, preemptive political and military actions, in order to prevent and localize conflicts, may have a decisive importance for the prevention of war. At this stage, besides political measures, important significance can be gained by economic sanctions, by sea, air and ground blockades of communication lines, by demonstrations of force, by the creation of peace-keeping forces for the disengagement of the conflicting sides, and other ways of actions. For these reasons, the demonstration of force against Haiti in September 1994 promoted the success of political and diplomatic actions.

In the event that these measures do not bring about a positive result, and when military actions seem to be inevitable, it is most important to ensure surprise actions, by a careful masking of the basic route of the intended action, and by effecting misinformation to the opponent.

Military actions of ground forces could be preceded by massive strikes of aviation and naval forces, with the purpose of breaking the enemy's will to resist. After or during fire strikes against the opponent, he can be presented with the ultimatum of surrender, or certain reconciliation conditions to the conflict can be offered. It is expedient to introduce major ground force groupings (which usually suffer the greatest losses), into what Clausewitz called bloody battles, only after the main fire deployments, and after the major enemy targets have been suppressed or destroyed. Taking into account the near-impossibility of putting out of action all means of hostile defense, it is essential, along with the destruction of major hostile targets in the zone of forthcoming operations, to concentrate the basic efforts of fire deployment for powerful destruction of the opponent, along the directions of attack of friendly troops, in order to create a sort of corridor in the enemy defense. When introducing ground force groupings into combat, it is better to avoid head-on, frontal actions against the most fortified defensive boundaries and regions, and swiftly to penetrate flanks and rear of hostile major groupings, creating fire corridors and bypassing the strongest units of enemy resistance.

However, even given the most skillful application of various ways of indirect actions, it will not always be possible to overcome the opponent. It will then be probably necessary to adjust and to limit political and strategic purposes, and should these appear impossible or lacking in positive results, it might be expedient to undertake direct actions. But in the future, even such actions will probably be carried out through well-prepared engagements and operations with skillful use of all the diverse ways of military operations.

Special methods of warfare, from psychological operations and subversive actions to operations of special task forces, will acquire a special place in the system of indirect actions.

In the USA and in a number of other countries, whole series of manual primers on the preparation and conduct of special operations have been developed. For example, the US Army Combat Regulations M31–32; *Control, Supervision and Support of Operations of Special Forces* (1981); US Army Regulations M31–32; *Techniques of Special Operations* (1983), and others.

Speaking generally, it should be mentioned that back in 1954 George Kennan came to the conclusion that the Soviet dilemma could not be solved by purely military means, and he called for the search for more flexible forms and methods of achieving the USA objectives in the international arena.[32] In the end these forms and methods of struggle against the Soviet Union and other countries of the Warsaw Pact turned out to be the most effective. The former US Defense Secretary L. Aspin spoke about the need to develop methods of intimidation of new and old enemies. But intimidation, in his opinion, should be based on some new hellish inventive principles.[33]

Some experts even believe that the world is entering into a new generation of wars, aimed not so much at the direct destruction of the opponent, as at the weakening of his military power from within, by way of political and economic pressures.

The American expert E.I. Keel formulates this purpose more subtly. He writes: 'We must develop an effective strategy, directed not at constraining our opponents, but at influencing them so as to shape the development of their foreign and military policy according to our interests.'[34]

The shape of future war presents some new aspects with regard to the scope of armed struggle. As we know, the twentieth century has witnessed an increase in the duration of wars (World War I lasted four years, World War II six), their spatial scope, and the number of states, populations and armed forces involved in war.

If in the future it will not be possible to avoid large-scale war, the tendency for the increase of these parameters might continue, but reverse tendencies may also reveal themselves. In the newly developing arrangement of military and political forces in the world, the overcoming of global confrontation and the threat of application of nuclear weapons will constrain any transition of war into world war. In the multipolar world, wars between individual states are the most likely. Therefore, numbers of states participating in war could be very diverse. This circumstance will influence the duration of wars. Although more powerful modern weapons can destroy the opponent in a shorter time, the economic possibilities of the restoration of losses, and the build-up of efforts during war, have also grown – as displayed in the Iran–Iraq war. Judging by the experience gained in the Gulf War of 1991, survival for troops and military targets has turned out to be greater than expected. Under certain circumstances even local wars can be prolonged, as in Vietnam and Afghanistan.

As before, large forces will participate in battles. Weapons and equipment will diversify and increase in numbers. In local wars, and more so in wars of greater scale, it is impossible to make do with professional armies only as in peace time, thus inevitably mobilization of all the forces and means of the state will be required. Therefore the proposition that 'nowadays wars are fought by the people' does not lose its relevance. Accordingly, it is still essential to prepare the population, industry, and the whole national economy for mobilization to carry out military objectives.

Principles of Military Art and Their New Contents

What then will change, what important transformations will take place in the future methods of the preparation and conduct of armed struggle? If one were to define them first in the most general and concise terms, the expected new phenomena could be reduced to the following:

First, the most essential changes will concern not only and not so much the external parameters of armed struggle, of which much has been said, although these will also experience some changes. The major changes rendering armed struggles of the future dissimilar to previous ones follow from its internal content, where the actions of different service branches and combat arms, carrying out a greater number of highly complicated, interconnected strategic and operational and tactical tasks will be compressed. At the same time, the actions of strategic systems, with ground, air and naval combats and battles, will affect the general course of military operations not only vertically (from strategy to tactics and vice versa), as in the past, but in multiple directions. The main objectives for destruction of the opponent will be carried out not during collisions of forward units, but mainly by application of fire from a distance. As a result, all combats and battles will acquire a dispersed volumetric character, embracing all spheres of military actions along the front, and in the depth and height. The intensity of fire power by all participants has been sharply increasing, causing new and probably extreme psychological pressures. The new character of future armed struggle will also arise from the internal density and dynamics of combat actions, and the general intensity of combat confrontation.

Second, the influence of weapons, especially nuclear weapons, on the determination of political and strategic objectives is growing. In particular, the role of strategic weapons as a decisive means of conducting war, directly ensuring the achievement of strategic objectives continues to increase.

Third, the spatial scope of armed struggle is expanding. The weapon systems of the future and the increasing combat potential of armed forces enable them to render powerful strikes against the territory of their opponents and against their military targets, carrying out not only consecutive, but also simultaneous destruction of their major groupings. During World War II the United States and some of the British Commonwealth countries, such as Canada and most of India, remained outside the combat zone. The same was also true during the Gulf War for the military bases of the multinational forces in Europe, the United States, and on several islands in the Indian Ocean. In future wars against strong, technically-equipped opponents, all bases and targets in even the most remote regions will be exposed to missile and aviation strikes. And in practice the concepts of 'front' and 'rear' will have a rather conditional character.

Fourth, the need to coordinate the efforts of all service branches and combat arms requires their joint application within the system of united strategic operations. The utilization of numerous diverse weapon systems and equipment in armed struggle makes future battles extremely complex and creates new conditions for their application and interaction.

Rapid developments in advanced technologies will increase the military and technical gap between the advanced leading states and other countries. Therefore, military art should assume the possibility of confrontation not only between opponents with relatively equal technological potentials but also between those with different levels of hardware.

Fifth, out of the three major elements of combat and battle – fire, strike and maneuver – greater importance will be attributed to fire destruction, which should better prepare for the strike, and increase the strike force without compelling troops, as in the past, to overcome the opponent at the price of large losses. Follow-up echelons and reserves will be destroyed, as a rule, even before they approach the battle position.

Decisive battles will occur not only on the ground and on the sea, but also in the air, and as a whole, operations and combat actions will have an air to ground character, when fire and electronic strikes by ground and air systems along the entire depth of the enemy deployment will be combined with repeated landings and penetration into the depth of the hostile defense by air-mobile units, delivering strikes not only from the front and flanks, but also from different directions in the enemy rear. As a whole, operation and combat actions will develop swiftly, without the presence of continuous fronts or only at their temporary stabilization, being highly maneuverable in character.

Considerable changes will occur in the character of the initial period of war, in methods of preparation and conduct of offensive and defensive operations, in the conduct of meeting engagements, in methods of fire destruction, in fulfillment of maneuver and the new approach to the creation of the necessary density of forces and means, with concentration of the basic efforts on decisive directions. Rapid and sharp changes in conditions, and the introduction of automated control systems will complicate and fundamentally transform the activity of commanders and staffs with regard to command and control of troops and naval forces. In the future, losses of personnel and combat equipment will be more significant.

With further consideration of new aspects of the armed struggle of the future, we shall place main emphasis on establishing continuity and new qualitative changes in the development of military art. But to complete the picture and for the sake of objectivity it seems necessary first to mention some other more 'radical' opinions. As we know, the appearance of the nuclear weapon presupposed significant changes in the ways of conduct of armed struggle, which in the literature were called a 'revolution in military affairs'. But this has also produced some extreme views, such as an attempt to present matters as if almost nothing remained over from the past theory of military art.

For instance, both the US doctrine 'of massive retaliation', and Soviet military theory (most fully explained in *Military Strategy*[35] which reflected the official opinion of Soviet military authorities) recognized that future world war will be general, nuclear-missile war of a transient, rapid character.

The basic form of strategic actions has to be nuclear missile strikes by strategic offensive forces, and the ground operations were to be reduced only to exploitation of the results of nuclear strikes. It was assumed that the nuclear weapon would open a way to unconditional and absolute victory over the opponent. At the same time, it was not taken into account that further qualitative improvements and quantitative accumulation of nuclear missile weapons will result in the extinction of 'nuclear strategy', oriented, in essence, on a revival of the *blitzkrieg* based on the use of superweapons. But super-radical extreme approaches have a tendency to merge and transform into their own opposites. Today, the very authors of these radical views already not only admit what has become obvious – that nuclear war is bluff, and any military actions in conditions of its application are impossible – but also expand this thesis on any war, including conventional war.

The chief of US DoD (Department of Defense) Office of Net Assessment, A. Marshall, believes that in the next three decades the present methods of the conduct of war of attrition will lose any meaning, and today's most modern combat equipment will lose its importance and become obsolete. In his opinion, 'the century of computer science' is fraught with a revolution in military affairs not less than the one produced in the fifteenth century by artillery, or by nuclear missile weapons in our time. Mass armies will give way to rather small formations of expert professionals who will be able to destroy the opponent without direct contact with it. Not only bayonet attacks, but also tank battles will be forgotten. Tanks, armored vehicles and cannon will be destroyed by smart homing missiles, launched from distances of hundreds of kilometers or from submarines. Aircraft carriers and other large ships will also be placed on the scrap heap. These will give way to ships of small displacement, built with 'stealth' technology, and armed with cruise missiles and unmanned aircraft. Proceeding from this, Marshall thinks that the Gulf War of 1991 is the last or one of the last armed operations of the industrial epoch. He believes that although it was fought with highly technological weapons, its methods were outdated.[36]

Some of these assumptions are not without foundation and probably correctly reflect tendencies in the development of military actions. But they miss the fact that changes in a complex network of military craft cannot occur only in separate links and in one direction, without touching upon others. Supposing that tanks, cannons and armored vehicles are immediately and totally destroyed from a distance, even before they approach the battle ground, then warring sides would be compelled to reject these means, like they rejected to use cavalry in times past, and would have to invent other means which possess greater survivability on the battlefield. Instead of modern tanks, APCs and cannon, there will be other means of conducting combat actions.

Even in the most remote future it will be impossible to fight a war with only long-range systems. One way or another there will be troops on the battlefield, so the possibility of collision between them is not excluded, although certainly in forms differing from previously existing ones.

Such extremists believe that any defense plans developed by general headquarters not only for nuclear, but also for conventional war, are useless. The only reasonable approach, they say, is to admit the impossibility of winning in any future war. Moreover, the largest drawback to all plans and scripts of war, is allegedly that they do not envisage the defeat of their own state. Military art should in the future adhere to 'non-offensive defense'; should refuse any offensive actions and attempts to occupy hostile territory. In addition, no attempt is made to answer the question: is it possible to free territory seized by an aggressor and to restore its sovereignty without conducting an offensive and by keeping within the bounds of defense? Yes. Such statements, regardless of their proponents' intentions, encourage surrender to potential aggressors, and deprive peaceful states of their potential for deterrence and prevention of war. Such ideology has nothing to do with interests of security, whether national or global.

But in reality war will go on as before, victories and defeats will occur as the Gulf War or the events in Caucasus and the Balkans have demonstrated. To think that to refuse to fight will stop war, or to assert that the refusal of victory in war is the surest way to a safer world, means to completely depart from real life and to live in the world of wishful thinking and imagination.

Some fashionable trends in recent military theory are presented as a swing away from traditional interpretations of categories of armed struggle. The authors of such views believe it is possible to decree 'admissible and inadmissible' methods of combat actions. Offensive methods of action are treated as admissible only in counterattacks, counterstrikes and in the transition to a counter-offensive. Troops in a counter-offensive have the responsibility to pursue the enemy only on friendly territory, but must stop at the border, not crossing over under any circumstances. It is suggested that even fire strikes on territory of the opponent should be limited. Thus, excluded from the process of combat actions is the phase of destruction of the opponent, which limits the concept of 'victory' within tactical frameworks, while the concept of victory on operational and strategic levels is abolished altogether.

The aspiration of some theorists 'to transform' military art from a humanist position is certainly noble. But unfortunately it is not viable, or even promising, for the development of military art is hardly influenced by wishful thinking, rather it is defined by objective factors.

Precisely this sort of theoreticians 'forever marching ahead of progress' have been talking about crisis and stalemate in military science. They mistake dead-ends of their own thinking for general crisis of military theory. In fact, the dilemma of the theory and practice of military art in the postwar years has been that ultra-radical theories have distracted it from studying and developing those rules and approaches which will really be necessary for future armed struggles.

Speaking of elements of continuity in the development of military art, it is necessary to emphasize that although there are no 'eternal' and immutable principles, there are certainly principles of military art, which have already survived through several millennia, and will probably survive for a long time ahead.

The fundamental principles of military art which are of long-term importance could be:

- formulation of feasible goals and objectives, corresponding to available means and resources;
- constant high combat readiness of troops;
- coordination and close interaction of all branches of the armed forces, the direction of their efforts towards achievement of the joint purpose;
- resolute concentration of efforts on major directions at culminating moments for fulfillment of the main objectives; maintenance of initial superiority in fire and radioelectronic suppression on the opponent;
- activity, perseverance and continuity of combat actions and fighting, with the utmost intensity necessary for the fulfillment of objectives;
- constant build-up of efforts during combat and operations;
- bold maneuver by troops, forces and means, evasion from frontal actions, the strive to advance into the flank and rear, and to make use of weak and vulnerable points in the enemy deployment;
- early creation of reserves, their skillful use and timely reconstitution;
- stratagem, deception of the opponent, surprise actions, the use of unexpected methods of achieving objectives; preemptive actions and disruption of the enemy's surprise actions;
- maintaining high morale among the friendly troops, and the reverse psychological influence on the opponent.
- all-round combat, material and technical support of operations and combat actions, maintenance and timely reconstitution of combat readiness of troops;
- efficient command control; uninterrupted reconnaissance and constant awareness of the situation; flexibility of plans, foresight and timely response to changes in the situation; firm and continuous control of troops during combat and operations, determination in achievement of formulated objectives, carrying out of the accepted decisions and set tasks;
- continuous support of troops and creation of favorable conditions

for fulfillment of the formulated objectives with the minimum of losses.

But as military art develops, these well-known principles will fill with new content and new meaning, while the forms and methods of their realization in the preparation and conduct of military actions will vary. There exists a fundamental difference between what a principle asserts, and the way it can be implemented in new conditions.

For instance, in the above-mentioned book, *Military Strategy*, it is said that the whole number of known principles, norms and rules, which previously guided military strategy, are nowadays exposed to a radical reassessment or even completely lose their importance. Among them the authors counted principles of concentration of forces and means on the main direction, economy of forces, and partial victory. The authors also believed that strategic deployment, strategic offense, strategic defense, strategic maneuver and other basic concepts of the theory of strategy have in many respects lost their importance as well.

This book rightly draws attention to the fact that long-held principle of concentration of forces and means on the main direction requires modification for new conditions. It should be implemented not by concentrating troops on an elected direction, but mainly by amassing means of fire destruction. But as a whole, this principle, and especially the skillful selection of direction of the main strike, and the concentration of forces and means on main directions, will remain important even in the future. However, methods of its realization will change. Modern long-range means sometimes allow maneuver by fire to deliver massive strikes to certain enemy groupings, instead of the maneuver and concentration of forces. The ground attack groups can advance in the necessary directions only at the last moment.

This also applies to strategic maneuver during war. The appearance of new means of destruction not only does not cancel, but even increases, the importance of strategic reserves and the necessity of their maneuver during war. However, issues of the dispersion of troops and naval forces, and measures to protect them from weapons of mass destruction cannot be solved by former methods and should be executed in light of the new requirements.

The outcome of any future war, to a considerably greater extent than before, will depend on the quantity and effectiveness of effort applied at the beginning of war. However, the strategic principle of economy of forces still in general applies, because a war between major states with massive potential will hardly be of short duration. It will be necessary to prepare for rather long, persistent, and fierce armed struggles.

In spite of the fact that powerful means of warfare are in the hands of the highest military command, and their application can render very considerable influence on the course of war, attaining directly strategic results, the principle of partial victory still does not lose its importance. The results of utilization of strategic strikes and destruction of hostile groupings will depend on the success of combat actions on the operational and tactical scale. And although the proportions of strategy, operational art and tactics have changed significantly, all these categories of military art retain their importance.

The Initial Period of War: Offense and Defense in the New Conditions

An important and difficult task is the forecasting of the probable actions of an aggressor in the initial period of war, as well as development of the appropriate ways of enhancement of combat readiness and strategic deployment of armed forces. These complex issues have to be dealt with anew each time, in view of radical changes in ways of conducting armed struggle on the eve of war.

Before World War II, the strategic deployment of armed forces was viewed as a complex of consistently planned and executed measures covering mobilization, concentration and deployment of armed forces in the theater of military action, which were carried out during the period of threat or at the start of war. Soviet military theory and practice committed serious miscalculations in this area in 1941. In *Military Strategy* these views were regarded as outdated. It was pointed out that the majority of these measures should and can be implemented beforehand, and completed during the period of tension prior to the outbreak of war.[37]

But this recommendation, although correct in theory, does not take into account the complexity of the task, and is much like an attempt by a fire brigade to arrive at the scene of a fire one hour before it breaks out.

First, this approach does not take into account the fact that aggressive states usually count on surprise assaults without preliminary deployment of all necessary forces and means.

Second, preliminary strategic deployment of armed forces prior to the beginning of war, despite all its benefits from the purely military aspect, is not always feasible for political reasons. Mobilization, let alone the whole complex of measures on a strategic deployment was always considered equivalent to a state of war, and reverting to a peaceful situation is difficult. It is not by chance that Marshal B.M. Shaposhnikov termed mobilization the 'opium of war'. If war in general is politics through and through, then on the eve and at the start of war its political aspects prevail even more strongly.

Taking everything into account, the system of strategic deployment cannot be guided unequivocally by only one of the most favorable variants, and should be more flexible, to provide an organized deployment of troops under any conditions of the initiation of war by aggressors.

In modern conditions it is especially important to maintain high combat readiness for the repulse of the opponent's sudden assaults. The perfection of means of assault and the stakes an aggressor places on preemptive strike, with the increased role of the time factor in the beginning of war, all set new requirements for the combat readiness of armies and naval forces. Any defensive character of the military doctrine not only does not lessen these requirements, but considerably increases them, because the aggressor, being prepared for war beforehand, chooses the moment of attack, and the defending party should always be ready to repel aggression. The complexity of the problem is also caused by the fact that the rapid assault of the opponent and his attempts to ensure a surprise blow, should be opposed not only with appropriate and effective means of protection, but also with a more advanced and flexible system of combat readiness of friendly armed forces. It should ensure the possibility of immediate reaction to the increase in danger, and the organized commitment of armed forces into action under the most difficult conditions.

Therefore, for armed forces to provide a fast reaction to any military conflict or aggressive actions their strategic composition should include: covering forces based forward, mobile forces, follow-up echelons and reserves. This system makes it unnecessary to place large forces prematurely on major strategic directions. Instead, it is possible to deploy the main forces at some depth, keeping them ready to be sent forward quickly on threatened directions.

In the future, the initial period will probably be the key and crucial period of war, in many respects predetermining its outcome. Unlike the past, war can now start without a ground invasion. Moreover, it can begin even before ground forces have completed concentration and deployed onto the theater of military actions.

War can now begin by a fairly long air operation or even campaign (consisting of a number of air operations), in which both air and naval forces render massive bomb, missile and radioelectronic strikes against the opponent's aviation, missile installations, naval forces, air defense system, control points, industrial and other major targets, and also later against major ground groupings. Both aviation and naval forces can accomplish this from distant bases, and also without a preliminary complete concentration in the theater of military actions. Aircraft and ships need only come as close as the cruise missile launch points. The latter can automatically find and destroy targets in any depth of enemy territory. As a result the whole country can turn into one continuous battlefield. All this will facilitate conditions for the enhancement of the factor of surprise.

Under cover of massive aviation and naval forces strikes, air movements and the movement of large combined-arms units will be accomplished. These units may switch to the offense only after the suppression of the opponent, sufficient to deprive him of the possibility of organized resistance. Such courses of action are also caused by the necessity to spare very expensive regular professional armies which are sensitive to heavy casualties.

Some military experts, engaged in forecasting the development of armed struggle, assume that in the future the application of ground forces, the seizure and occupation of enemy territory will not be required. Such a variant cannot be excluded in cases when as a result of powerful air operations the opponent is completely broken, and surrenders. But in the majority of cases in a war against powerful opponents the termination of war without the use of ground forces is hardly conceivable.

The Gulf War is a vivid example. Iraq formally surrendered, but without the occupation of its territory by ground forces. Many military and political ends, for the sake of which the US and its allies started the war, had not been attained. Only the intrusion of ground forces, supported by the air force and navy can consolidate success and make further resistance impossible.

In the course of a war against a powerful and active opponent it is hardly possible to expect to win by air strikes only. The destruction of large enemy groupings might take lengthy and intense efforts, and requires the execution of a number of consecutive operations and combination of various ways of armed struggle. In this connection, such future forms of strategic actions as the strategic offense and strategic defense, with regard to new ways of their preparation and conduct, will certainly not lose importance.

In modern times there are no states with military doctrines openly proclaiming aggressive, offensive goals (as Germany or Japan did in the 1930s). The majority of countries one way or another emphasize their disinterest in war, in other words, a defensive character of military doctrine in the military–political aspect. But this does not mean that the actions of armed forces during war only have a defensive character.

Because of deep-rooted false stereotypes in Russia, the CIS countries, and sometimes in the western press around all aspects of military doctrine, the most discussed issues are the actions of armed forces within the defensive character of military doctrine. Some rather conventional concepts – 'offensive' or 'defensive' war, 'offensive' or 'defensive' doctrine – have always characterized only the aggressiveness or non-aggressiveness of the state, determining the ways of actions of armed forces only in the very beginning of war. In the course of war both parties inevitably resorted to offensive, as well as defensive actions. On the Soviet–German front during World War II the Wehrmacht during the first two years of war was mainly on the offensive, and subsequently on the defensive; the Soviet army on the contrary, was in the beginning compelled to defend and then in the second half of the war conducted mainly offensive operations.

Once war has been unleashed, even the most peaceful state will submit its policy to the conduct of war and the protection of its territory. Any army, in order to repel aggression and protect the country against fatal danger, will act as required by a specific situation, proceeding first of all from operational-strategic expediency, and using the whole arsenal of military art. Therefore, recent suggestions made in Russia on the necessity to determine by law the nature of military action that an army should undertake during war seem naïve. The defensive interests of a peaceful country require that troops be trained for both defensive and offensive operations, as is done in NATO, China and other countries.

Even with a defensive military doctrine, basically directed on retaliatory, defensive actions in the beginning of war, it is impossible to exclude a variant, when in the case of an apparent threat of aggression (provided the reconnaissance is well organized and the decision for strategic deployment was taken in advance) immediate transition to the counter-offensive might be undertaken with the beginning of enemy aggression.

One has, of course, to take into account that in the beginning of war it is probable that a peaceful country will initially conduct defensive operations. Therefore, alongside the elaboration of offensive actions, in modern conditions more thorough preparation of the organs of control of troops to conduct defensive operations is required.

It is well known that the Soviet army paid dearly in 1941–42 for its underestimation of defense. The experience of the battle of Kursk has convincingly shown the legitimacy, under certain conditions, of defense on the strategic scale. This battle introduced a new point into military art, because the Soviet troops had reverted to defense not out of necessity, because of lack of forces and means (as should have been done in compliance with existing theoretical views), but intentionally, having forces superior to the opponent. However, the idea was later expressed (by Marshal M.V. Zakharov) that the 'defense of Kursk cannot be viewed as a classical, typical case, as a sample for imitation'.[38] This is one of the examples where new phenomena, born out of the experience of war, were brought in to suit previously existing theoretical notions.

In theory, it was for a long time considered, and still is, that to repel an enemy offensive it is enough to have two to three times less forces than for an offensive operation. But the experience of World War II did not confirm this notion. During that war there was not a single successful defensive operation conducted with considerably smaller forces than those of the attacking enemy. Certainly, when forces and means are not sufficient, there is no other way but to revert to defense and to hold the line for some time.

But if the opponent mounts an offensive, it is not always feasible to stop him with a small force, let alone frustrate an offensive and defeat the attacking opponent. In 1941–42 it was only possible to stop an enemy offensive after a long retreat for hundreds of kilometers, and commitment of fresh reserves into combat. It is possible to repulse attacks of superior enemy forces in defense at the tactical level. But on the operational-strategic level, given powerful means of fire destruction and high maneuverability of troops and aviation, the attacker, holding the initiative, has the chance to create a multiple, overwhelming superiority on selected directions, and it will take rather large forces to withstand his deep penetrations. Therefore, as war experience shows, the conduct of defensive operations, when planned for the successful repulsion and frustration of an extensive hostile offensive, requires as many forces and means as an offensive. Liddell Hart stressed this point when proving that the forces needed for a successful NATO defense should be almost equal to the forces of the attacker.[39]

When forces and means are not sufficient for defense, and you are compelled to revert to defense against a superior enemy, it is probably expedient to conduct a maneuver mobile defense, fighting back along intermediate lines until the arrival of large follow-up echelons and reserves. When the situation demands it, it is better to use a pre-planned, organized, controlled withdrawal, than to try in an obviously hopeless situation and at any cost to keep each defensive position, and to doom the troops to a forced, unorganized retreat. Such a retreat, despite all severe repressive measures, usually causes chaos, panic and disorganized actions, resulting in more complex operational-strategic consequences than an organized and authorized withdrawal.

In analyzing the problem of offensive and defensive operations, it is necessary to take into account the tendency towards further blurring of the distinction between the actions of troops on the offense and in defense which is now more and more clearly displayed. What is more, a modern offensive stands for a combination of fire strikes, the rapid movement of tanks and armored infantry supported by aviation and combat helicopters from the air, dauntless actions of air assault forces on the flanks of confronting groupings in the depth of defense. The modern offensive is not a successive movement of troops from one line to another, but more of a resolute simultaneous destruction of the opponent throughout the depth of his formation.

With the increased range of tactical missile systems and artillery, there appears the need to specify ways of dispersion of troops along the front and in depth, both in offense and in defense, and to move from assembly areas, deployment lines, and reserve echelons. It is very important to destroy hostile missiles, nuclear facilities and precision weapon systems prior to the advance of major groupings of the troops into assembly area for offensive and for launching counterstrikes.

Fire and maneuver in combination with the reliable maintenance of occupied positions will constitute the basics of defense. It will be characterized by the application of powerful fire means, to inflict a defeat on superior enemy forces even from a long distance away, backed up with a high level of activity of the defending troops.

Encirclement operations will also develop in a different way. In World War II, encirclement was mainly carried out by way of the rapid cutting of mobile troops into the flanks and rear of enemy low-mobility infantry groupings. Under the new conditions highly-mobile encircled troops will have increased chances for maneuver, the formation of attack groupings, and breakout from encirclement. At the same time, the dependence of material factors on the supply of petroleum, oil, lubricants, ammunitions and the availability and quality of roads is increasing. This is a drawback for encircled forces. Therefore it is essential to find ways of a more reliable and rapid execution of this complex form of operation in light of the new conditions.

Tactics will develop especially quickly. Tactics are the most dynamic area of military art. Each passage from one generation of weapons to another is first of all reflected in tactics, bringing to life new tactical methods of warfare, and the specification or reorganization of the composition of units and methods of their training. With ground forces, for example, it is essential to elaborate forms of interaction of tanks, infantry, artillery and aviation; ways to suppress the antitank defense of the opponent; methods of dealing with remote mine laying; airmobile antitank means, army aviation and so on. In each of the services there are many problems in the field of preparation and conduct of modern combat on the ground, in the air and on the sea.

The growing importance of long-range combat is a characteristic feature of tactics. Long-range fire enables troops to inflict significant losses on the opponent long before direct contact. Moreover, with an increase in the missile launching range, and the firing range of artillery, depth of fire combat also increases. Modern weapons allow a seamless offensive, surprise and rapidity of strikes, high activity and stable defense.

As the actions of ground forces, especially in the beginning of war, could in a number of cases be preceded by protracted and massive air operations and campaigns, the role of aviation is increasing significantly, while the repulsion of enemy air strikes, and the struggle for domination of the air cease to be missions for the airforce and air defense only. They become the main purpose and basic content of combined-arms combats and operations. Struggles against air opponents should be conducted with all available forces, including missile forces and artillery, air assault forces, subversive groups and other means. This requires the complete subordination of all forces and means, including air force and air defense facilities, to the commanders organizing the combined-arms combat and operation.

In the days when aircraft flew directly over target regions to deliver strikes, and dived on columns and combat orders of troops, it was important that each unit and each element of the combat order had its assigned anti-aircraft means. Nowadays, when aviation can strike without entering an air defense zone and without needing to approach targets closely, the organization of air defense will gradually lose its importance, because the allocation of regular air defense means to each unit would be prodigal and irrational. A centralized zonal system of air defense, first at the level of units (divisions), and then at the level of large units (corps, armies) will probably be the most expedient. Such systems will also facilitate the notification of air defenses on flights of friendly aircraft, and will exclude their destruction by anti-aircraft facilities.

The introduction of rocket-assisted projectiles in artillery led to a two-fold increase of range and rate of fire, enabling destruction of the enemy to a depth of 25–30 km beyond its forward area. Artillery power is growing too. Thus, in the last 10–15 years the number of artillery systems of 120 mm caliber and above has increased five or six times. The development in the US Army of the rounds 'Copperhead' and 'Staff' and the automated system 'Tacfire' connected with reconnaissance means, have increased the accuracy of artillery fire and flexibility of control. Suffice it to say, that topogeodesic locating of the battery takes five minutes (it is planned to reduce this to 2–3 minutes), planning of fire for a battery against 35 targets takes 109 seconds. It takes 1.5–2 minutes for a concentration of fire of several batteries, and for planning division artillery fire 15 minutes (in the near future planning time will be reduced to eight or ten minutes). Not so long ago this task required from two to six hours to perform.

During World War II the readiness of troops to assume the offensive was mainly limited by the time necessary for preparation of artillery. Modern capabilities of artillery greatly reduce preparation time for offensive combats and operations. The secrecy of preparation for an offensive has increased, because it became possible to concentrate attack groupings on selected directions only at the last moment.

Tank troops require particular mention.

In the early days, tanks were conceived as a means of assistance to infantry in the penetration of defense. But after they were formed into combined arms units, and formations with infantry, artillery and other arms, tanks, as a combat weapon, acquired operational importance. The application of tanks in close coordination with other arms, especially supported by army aviation and under a cover of new air defense means, opens up new possibilities.

It is argued sometimes that in due course tanks will become obsolete. This was much discussed even before World War II when tanks dominated the battlefield. In the USSR N.S. Khrushchev 'buried' tanks after the appearance of the nuclear weapon. However, these assessments have not been substantiated. The actual tendency is such that not only the tanks themselves as a means of increased survival and mobility of heavy attacking weapons, but also a means of combat support (reconnaissance, engineering means, control posts, self-propelled artillery and others), have become armored and mobile.

The appearance of new material significantly raises the defensive capabilities of troops. The penetration of defense is complicated by the increased fire power of defending troops and wide introduction among troops of precision antitank missiles with greater range of fire. Their maximum range has increased five or six times, and their performance against armor 2.5 to 3 times. The probability level of the destruction of tanks by prospective antitank means can increase from 0.7 to 0.8. Combat helicopters and assault aviation are especially effective for the destruction of tanks and other armored vehicles. At the same time the quantity of antitank means is being further augmented. For example, a modern American division acting along a front of up to 30 km can create a density of antitank means of up to 30–35 units per 1 km of front.

New systems of ground (artillery, rocket-assisted) and air remote mine laying at distances of tens and hundreds of kilometers have been developed. Extensive employment of mine obstacles will complicate the use of tanks.

The appearance of reconnaissance-attack complexes, connected with the missile systems 'JTACMS', and also 'PLSS' have created conditions for the destruction of tanks and radio-signalling systems not only directly on the battlefield, but also in depth for up to 200–250 km, thus complicating the advance of troops, especially of the reserves.

But the material of warfare is being constantly improved. Tanks now possess increased fire power, reinforced armor protection and maneuverability, while reduced or only slightly increased in weight. Gun caliber has increased 1.5 times; the firing range is enhanced; tanks are equipped with combination gun-launchers. Better surveillance and target acquisition devices and armament stabilization systems enable more accurate fire on the move. The performance of tanks against armor has increased thanks to gains in the fire power of guns, and the efficiency of high velocity armor-piercing and hollow charge projectiles. Ammunition with combustible cartridges is also used.

The increase of armor protection for tanks and other kinds of armored vehicles is due, not only to an increase of thickness of the plates, but also through the use of stronger synthetic materials, methods of passive and active protection, installation nets, screens, and other facilities over the armored vehicles, causing premature explosion of shells, radioelectronic systems, shell deflection, and so on.

Most modern tanks M1A1, M1A2 ('Abrams'), 'Leopard-2' or 'Leclerc' and T-80 are nearly equal in maneuverability, fire power and armor protection. They are equipped with more sophisticated weapon control devices (laser range-finders, computer devices for preparation of data for fire and provision of accuracy of aiming).

One of the ways to ensure the survival of tanks is to reduce their dimensions. For example, the probability of hitting the Swedish-made 'S' turretless tank decreases by 30 per cent. In the prototype US tank XM-803, with powerful hydro-pneumatic suspension and variable clearance of 200–600 mm the crew can carry out surveillance from the tank with a periscope, without leaving shelter, and lifting the turret only for fire. All this is especially important in the application of tanks in defense.

The survival rate for tanks is also improved through greater maneuverability, and by equipping them with more powerful engines with a specific capacity of up to 25 hp/t and more, in order to increase their average speed and crosscountry mobility. More sophisticated transmissions, driving controls and suspension systems are being developed. 'Flying' or 'jumping' tank projects have been mentioned in the press. In the age of technological progress it is impossible to exclude this possibility, but if tanks begin to fly, they will probably shift to the category of combat helicopters or other flying vehicles. The appearance of air-cushioned APCs and light tanks is conceivable. Robotics (robotank, APCs, self-propelled artillery cannons, robots equipped with machine guns and tank weapons) can play a certain role in overcoming hostile obstacles and zones of strong radioactive and chemical contamination. These combat robots will constitute an attacking echelon, which can be followed by conventional infantry and tanks. Thus, we evidence the enhancement of capabilities not only of anti-tank means, but also the combat possibilities of tanks in the struggle against them.

Effective actions by tanks and APCs will be conditioned by the constant support of aviation, primarily helicopters.

In general, in local wars against poorly equipped opponents, precision weapons and other new combat means display their superior effectiveness; but in a struggle with an equal opponent when various means are extensively used and both sides are under the pressure of active fire and radioelectronic warfare, the effectiveness of modern weapons systems becomes to some degree leveled out, and turns out not to be as high as is often represented by results of firing range tests.

The following example confirms this. The application by US troops of the anti-radar homing missiles 'Shrike' in Vietnam proved their superior effectiveness. In the beginning each first or second missile, and subsequently each third or fourth one destroyed enemy radar stations. But when on 18 April 1971 the Israeli airforce had for the first time applied 72 'Shrike' missiles against Egyptian troops in the Suez canal zone, only one missile hit radar station P-35 in the region Abu-Suveier, while other missiles missed their targets. Myself, then serving as chief of staff to the chief military adviser in Egypt, together with Egyptian officers and in particular the chief of staff of the Egyptian airforce General Hosni Mubarek (now president of Egypt), had to survey carefully the reasons for such low effectiveness of the 'Shrike' employment.

Analysis has shown, that in Vietnam an object-oriented system of air defense was used, with separate radar stations widely spaced. In the Egyptian canal zone along the 120 km front, there were more than 100 anti-aircraft units, each with their own radar stations, in addition to other radar stations. When all of them were in operation, their radar emissions affected missiles in such a way that they acquired different emissions and in the end lost their targets. Extensive application and especially active counteraction with other new means can lead to similar situations.

Enhanced fire power from ground troops and aviation, and their massive application, can result in the destruction and serious weakening not only of separate units and objects, but also of whole hostile groupings, forming breaks in the enemy's combat formation, and facilitating the fast penetration of attacking troops in depth. In combination with a high level of troop mobility, their continuous support by aviation combined-arms operations will have a highly mobile nature.

The assumption is sometimes expressed that in future armed struggle there will be no need to penetrate hostile defenses at all. But both the First and Second World Wars began with maneuver combat actions; subsequently each side tried to close gaps, to prevent penetration by enemy groupings in the friendly rear; as a result the front was stabilized for some time. Without a relatively stable front line it is impossible to ensure communication lines, supply of troops, combat activity of missile forces, aviation, air defense facilities and control posts deployed in depth. Therefore even in future wars, although less often than before, a front line will be relatively stable from time to time and that will require its penetration. In general though, combat actions will certainly have an incomparably greater maneuverable nature than in the past.

It is probably not justified to think that military actions can be conducted continuously and with identical intensity on all directions and throughout the course of the whole war. Not only in protracted, but even in rather short-term wars, there are operational pauses caused by the balance of forces of the opponents, the necessity of replenishing losses and materiel or other reasons that emerged in the course of combat.

Hardly viable are speculations that the availability of reconnaissance-striking complexes enables the immediate destruction of major hostile fire means and objects as soon as they are detected, and hence in the course of war there will be no need to conduct air operations and massive air and missile-artillery preparations, centralized at the levels of front (groups of armies), and armies (groups of corps), prior to the beginning of an operation or in the course of it.

First, when it is necessary to penetrate a defense or to break organized resistance of the opponent on a certain line, massive fire destruction of the opponent will be indispensable. Second, it should be taken into account that during war, under all circumstances, the quantity of ammunition supplies will be limited. During World War II, especially with the transition to defense, the quantity of artillery shells and air bombs was strictly rationed. In future war, guided missiles, bombs, artillery shells and other kinds of ammunition for precision weapons might be in especially short supply, because they are rather expensive, and disruption of communication lines, factories and warehouses will complicate the production and delivery of munitions. Therefore, there is no reasons to believe that troops will have an unlimited possibility to use expensive munitions. And it will not always be possible immediately to destroy each detected target. Hence, target points will sometimes pile up and for their destruction it will be necessary to carry out a massive fire destruction, provided the necessary means and ammunitions are accumulated, and as the situation demands.

On the Air–Land Operations

Aviation plays an important role in the conduct of any operations, especially at the initial period of war. This is because there are increased possibilities for aviation in conducting missile-bomb strikes against hostile targets, especially mobile targets, without entering the air defense zone; on the interception and destruction of hostile aircraft in the air, the defeat of aviation on airfields, and the launching of massive strikes against large enemy groupings, his fire and radioelectronic systems. Not only tactical aviation is now equipped with precision conventional weapons and all-weather means of destruction. The heavy bombers B-1B and B-2 have been modernized so as to enable them to apply precision non-nuclear means of destruction.

During the Gulf War, for strikes against Iraqi infantry and tank divisions, some 300–400 tactical aviation sorties, and 120–140 combat helicopter sorties were carried out, resulting in the destruction of 35–45 per cent of personnel and equipment.

Increased possibilities of transport aviation and helicopters, and the creation of special airmobile troops, necessary for rapid maneuver by troops and for the overcoming of obstacles and contaminated areas, will result in more frequent and massive movements of troops by air.

In this connection, operations will develop on the ground and in the air with simultaneous actions against the opponent at great depth. In short, they will have an air–land nature. In Soviet military art combined-arms operations were always considered in such a light; they used to be called ‘deep operations’, but from the 1970s they were called ‘three dimensional operations’. In the US army and other NATO armies the term ‘air–land operation’ is used. This is not some separate operation, but an operation of groups of armies, or fronts, expressing their new spirit and new peculiarities by an organic connection of all forces and means, acting on land, in the air and in the ether. Such an operation assumes the presence of ground and air echelons (air assault forces, airmobile units, aviation).

In military literature, the air–land operation is often mentioned as something that has already had its day, and which almost relates to past military art. But, in fact, the air–land operation is only just arising and is an operation of the future. Even many advanced armies do not have all the necessary materiel to realize the concept of air–land operations in their totality. Such operations have not yet been tested in war against strong technically equipped opponents, who are also capable of acting according to the principles of air–land operations.

Air–land operations presuppose: fire and radioelectronic defeat of enemy troops at all depths of deployment of his army, aviation and combat helicopters; division and precise determination of zones of responsibility for conduct of reconnaissance and destruction of enemy targets for formations and large formations; highly maneuverable actions of combined-arms units, swift offensives in flank directions and in depth, with the aim of dismembering and destroying the enemy in detail; multiple landings of tactical and operational assault forces in the enemy rear; dispatch of special operations groups to assist forward units of ground troops in forcing rivers and seizure of strongholds and airfields; encirclement of enemy groupings, disruption of his communication lines, delay of advancement of follow-up echelon and reserves, the destruction of nuclear missile weapons, control posts and precision weapons complexes. The essence of the air–land operation is an active struggle for the initiative, domination on land, and the achievement of air predominance. Air predominance is achieved by the way of the destruction of hostile aircraft and helicopters on the ground and in the air, suppression of the enemy’s air defense system, and a reliable air defense system protecting friendly forces. Special importance is given to the struggle against enemy follow-on echelons and reserves, with the aim of delaying their advance and defeating them before they are committed to battle.

The advance and commitment of follow-on echelons and reserves will present a difficult task for the attacking side too. Their vulnerability to strikes by aviation and precision weapons of the ground forces requires that they be not deployed to attack positions as large formations (armies, corps). It will probably be necessary to commit them piecemeal, or more likely to advance them from different directions and to merge units and formations for joint strikes only at the moment of attack.

New satellite-supported navigation devices, and more sophisticated night vision means, provide for the continuous conduct of combat actions by day and night, as well as in conditions of low visibility and adverse weather.

Thus, we witness the emergence of completely new defense problems, which should now be able to resist an enemy offensive by deflecting his massive strikes on land, from the air and other directions. To this end, defensive operations need to be far more of an air–land nature.

In a defensive operation, the main point is preemptive fire and radioelectronic strikes on the opponent, on the ground and in the air, with the aim to frustrate his offensive, to isolate the battlefield and defeat follow-on echelons and reserves before they are committed to battle; to conduct counterstrikes and counterattacks with landings of air assault forces and other airmobile units in the enemy rear; and the struggle against airborne assault echelons of the opponent.

In consideration of this, each element of a combat formation (including control points, artillery in positions, rear units) should be supported with the necessary means, prepared for circular ground and air defense, especially for the struggle against tanks, which may penetrate deep into defenses along with air assault forces and special operations groups.

The main elements of a defense system should be firepower and mine obstacles. Just as before, it will be indispensable to organize deep defensive lines and positions with a system of strongholds and centers of resistance connected by fire interaction, adapted for circular defense. At the same time, it is imperative to create more severing lines and positions in order to interdict and hamper enemy maneuvers to gain the flanks and rear of the main friendly groupings.

Defensive activities will be affected first of all thanks to effective fire defeat of the opponent, and maneuver by forces and means with the aim of localizing hostile penetrations. Provided conditions are favorable, the defending side can carry out counterattacks and counterstrikes. But it is necessary to take into account the fact that, given huge firepower superiority and air supremacy of the opponent, counterattacking troops can suffer heavy losses even before they secure a line to mount counterattacks and counterstrikes. Therefore, counterattacks and counterstrikes should not be ends in themselves, and in a number of cases it is more favorable to use follow-on echelons and reserves to repulse the enemy offensive from prepared positions, with a subsequent transition to the counter-offensive. In any case, the major conditions of the successful realization of a counterattack will be: skillful timing of the counterattack or counterstrike, when the enemy grouping that has penetrated friendly defenses is in the most vulnerable situation; secret preparation of the counterattack (counterstrike) and transmitting misinformation to the opponent concerning the line of advance of follow-up echelons and reserves; provision of firepower and air superiority, at least for a short period of time, on the counterstrike direction; reliable fire strikes on the opponent and air defense covering of counterattacking groups; swift actions of troops in the conduct of counterattacks and counterstrikes.

Transition to the counter-offensive should be carried out in conditions when the main offensive groupings of the opponent are weakened, and he needs time to organize and introduce new reserves. Attaining domination of the air before the counterstrike is especially important. Transition to the counter-offensive without preliminary stabilization of the situation in the course of ongoing combat actions, and without waiting for the termination of concentrated attack groupings, will probably be most characteristic.

In the beginning, air and fire operations aimed at the destruction of major enemy groupings can be carried out, and under their cover, attack forces might be rearranged and advanced. Transition to the counter-offensive by different units might not be executed in a number of cases, at different times depending on the achievement of the turning point, along one or another directions.

In defense as well as in offense, units and formations will be assigned wider areas of action than before. What is more, if in defense the successful rebuff of a large enemy offensive requires forces equal to that of the attacker, then on the main directions the width of division frontage, as the basic operational unit, can be about identical both for offense and defense, and could reach up to 20 km, or on secondary directions up to 30 or 40 km. This is effected mainly by increasing the combat power of units, and their ability to create the required densities of firepower without concentrating troops on narrow areas of the front, as well as by allowing greater freedom of maneuver for troops.

Significant Peculiarities of Combat Actions in Local Wars and Military Conflicts

In local wars and military conflicts a prevailing importance is attributed to social and political aspects. Social and political measures, supported by the majority of the population, should be the leading deciding factors for the settlement of conflicts, especially where national-liberation movements and civil wars are concerned. The interference of external forces in the internal affairs of other peoples and states, or the unequivocal support of only one of the conflicting sides, will bring no results. Settlement of conflicts is only possible with due respect to the interests of and thorough reconciliation of the opposing parties.

Experiences of war in Vietnam, Afghanistan, the Middle East, and recently in Yugoslavia, Somalia and other regions, clearly show that application of military force without proper political measures of settlement, severe bombings and repressive actions against not only armed formations but also the civilian population, only aggravate and expand the conflicts, and give them a protracted and sometimes even irresolvable nature.

From the military point of view, combat actions even in future conflicts will have a focal character. The front line will not always be precisely outlined, and actions by troops will be complicated by the mixing of armed formations with the local population. Operations of hostile aviation will occur from unexpected directions. The supply of troops, and all kinds of combat material and technical support, will sometimes have to be carried out in conditions of broken lines of communication. Precise organization of the control and the cooperation of covering forces and deploying mobile forces, as well as army groupings participating in combat action units of the airforce, air defense and naval forces, is required.

Problems of Troop Survival

In all kinds of combat actions, sustaining the survival of troops under heavy fire directed by the opponent against advancing units and troops located at depth, presents a serious problem. A higher survival rate for personnel and equipment can be attained mainly through mounting crews of all arms on armored vehicles, by the application of existing and new camouflage facilities, by covering equipment with special structures and paints that complicate detection by aircraft and helicopters, by radioelectronic and other means of counteracting enemy systems. But the most reliable and effective way to ensure troop survival and to shelter personnel and equipment remains the way tested through centuries – the skillful exploitation of natural protection feaures, the digging and thorough preparation of the ground. Experience from recent local wars, including the Gulf War, has convincingly shown how difficult it is to hit underground sheltered and periodically moving missile launchers, artillery tubes, tanks and other equipment. In any case, troops show higher levels of survival than are assumed from the basis of theoretical estimates and results of firing range tests. It is reported that American aviation in Vietnam dropped more bombs than during the whole course of World War II, but that the effectiveness of their application was low.

New earthmoving machines and self-digging equipment mounted on combat vehicles will facilitate faster preparation of the ground.

One of the most urgent and practically unsolved technical problems in this area, is the development of engineering machines to ensure productive work in winter, in deep frozen ground or rocky areas.

Radar, radio-relay and radio stations, towering above the surface, remain vulnerable installations. But current research and development allows the hope that these means, in due course, will become more portable and more sophisticated structually.

On the Issue of Losses During Armed Struggle

As more powerful means of destruction appear, and armed struggle becomes increasingly complicated, war attains a steadily more annihilative and destructive character. The number of victims among civilians grows steadily. It is a known fact that in World War I more than 10 million perished, 95 per cent being servicemen and five per cent the civilian population.

But in World War II 50 million people were killed: 52 per cent servicemen, and 48 per cent peaceful inhabitants. In local wars and conflicts from 1945 to 1982, losses have amounted to 35 million people, of whom more than 30 million were civilians. During the Korean war from 1950 to 1953, 84 per cent among the nine million killed were peaceful inhabitants. During the war in Indo-China in 1964–1972, civilian losses made up 98 per cent of all losses, while in Afghanistan they made up 85 per cent of all losses. Such dreadful casualties are explained not only by the escalating destructive properties of weapons, but mainly by their barbaric application. Frequently, in the so-called small-intensity conflicts, more people perish than in combat actions of regular armies. And only recently, as a result of the ethnic conflicts on the territory of the former USSR, ten times more people have been lost than losses by Soviet troops in the course of the whole of the Afghanistan war.

Taking all this into account, the necessity of adopting stricter international norms and agreements for the preservation of a peaceful population, has recently been persistently put forward for discussion. A complex of measures aimed at uniting the efforts of the world community in support of the effective protection of victims of military conflicts is needed. It is proposed to adjust the system of monitoring and the rapid reaction of the international community, with adequate means, for infringements of military legality. The creation of effective international justice, and an 'international office of prosecution' is also urgent.

In this case, hoping to avoid nuclear war, we do not touch upon well-known forecasts about possible catastrophic losses, which could be caused by the application of weapons of mass destruction.

Considering the nature of armed struggle, it is also necessary to take into account the growing losses of personnel and equipment. In the course of World War II the deaths of Soviet troops amounted to 15–30 per cent from immediate army operations, and 70–85 per cent in medical facilities. As shown by the experience of recent local wars and appropriately-transformed estimates, application of modern precision weapons can increase irrevocable losses to 25–35 per cent. Medical losses will increase accordingly. Losses of equipment could increase 1.5–2 times. There is no need to talk about losses of troops in nuclear war.

Obviously, the replacement of losses will present a difficult and rather important problem. The experience of the wars in the Middle East and the Persian Gulf region has shown that losses in combat, even with application of conventional weapons only, will increase significantly, because the effectiveness of the means of destruction rises sharply. In this connection there is a need to accumulate during peacetime trained mobilization resources to develop ways to amplify efforts and create new formations during war, replenish troops, repair and restore equipment, organize and use reserves, replenish ammunition stocks, petroleum and oils and other material, in conditions of increased consumption and mass destruction of lines of communications.

NOTES

1. S.S. Birusev, *Problemy revolutsii v voyennom dele* (Moscow, 1965) p. 35.
2. Cited from: P. Crosser, *Dialektika voyennoy tekhniki i ee posledstviya* (Moscow: Progress, 1975) pp. 91, 92.
3. H. Kahn, *On Thermonuclear War* (Princeton, 1960) pp. 22, 42.
4. B. Goldwater, *Why not Victory?* (New York, 1960) p. 22.
5. E. Knoll in, *The New York Review of Books* (Feb. 1984) p. 34.
6. *Izvestiya* (2 June 1994).
7. P. Crosser, op. cit., p. 198.
8. J. Bernal, *Mir bez voyny* (Moscow, 1960) pp. 21–6.
9. A.A. Grechko, *Na strazhe mira i stroitel'stva kommunisma* (Moscow, 1971) p. 55.
10. Cited from: D.M. Proektor, *Mirovye voyni i sud'by chelovechestva* (Moscow, 1986) p. 226.
11. For example: B.H. Liddell Hart, *Deterrence or Defense* (1961); R.J. Osgood, *Limited War* (1957); H. Kissinger, *Nuclear Weapons and Foreign Policy* (1959); B. Brodie, *Strategy in the Nuclear Age* (1959); M. Taylor, *Nenadezhnaya Strategiya* (1960); E.J. Kingstone-McKlorey, *Global Strategy* (1959).
12. M. Taylor, ibid., pp. 35–6.
13. Gaywin in, *Ordinance*, No. 3–4, 1958.
14. N.V. Ogarkov, *Istoriya uchit bditel'nosti* (Moscow, 1985) p. 51.
15. *Krasnaya zvezda* (23 Sept. 1994).
16. See more in: *Krasnaya zvezda* (24 Sept. 1993).

17. *Za rubezhom*, No. 23 (1994).
18. *Krasnaya zvezda* (8 Nov. 1993).
19. V.V. Stefashin, *Voyennaya nauka v Kitaye* (manuscript) pp. 126–35.
20. *Rossiiskie vesti* (4 August 1994).
21. V.V. Terekhov *et al.*, *Professional'naya armiya za rubezhom. Istoriya, opyt i usloviya sozdaniya* (Moscow: Institute of Military History of the Russian Federation, 1993) pp. 25–6.
22. Ibid., p. 159.
23. *Mezhdunarodnaya zhizn'*, No. 7–8 (1994) pp. 12–13.
24. *Krasnaya zvezda* (24 Nov. 1993).
25. *Izvestiya* (4 Nov. 1993).
26. *Krasnaya zvezda* (14 Jan. 1994).
27. L.S. Semeiko, in *SShA: Ekonomika, politika, ideologiya*, No. 11 (1992) pp. 33, 35.
28. *Za rubezhom*, No. 45 (1993).
29. K. Clausewitz, *O Voyne* (Moscow: Voenizdat NKO, 1936) pp. 67, 70, 71, 75.
30. A.A. Svechin, *Strategiya* (Moscow, 1927) p. 151.
31. B.H. Liddell Hart, *Strategiya Nepryamykh deistvii* (Moscow, 1957) pp. 444, 470.
32. G. Kennan, *Realities of American Foreign Policy* (Princeton, 1954) p. 81.
33. *Izvestiya* (3 Nov. 1993).
34. Earl E. Keel, Jr. 'Toward a New National Defense Strategy: The Need for Immediacy', in: Mary A. Sommerville, ed., *Essays on Strategy*, X, (Washington, DC: National Defense University Press, 1993) pp. 45–73.
35. V.V. Sokolovsky (ed.), *Voyennaya strategiya* (Moscow, 1963).
36. *Kompas*, No. 119, pp. 3–7.
37. *Voyennaya strategiya*, op. cit., p.22.
38. M.V. Zakharov, *Kurskaya bitva* (Moscow: Nauka, 1970) p. 454.
39. B.H. Liddell Hart, *Deterrence or Defense?* (1962) pp. 124–31.

CHAPTER 4

PROSPECTS FOR MILITARY CONSTRUCTION

> The defense of a country is equated to the value a nation puts on its sovereignty.
>
> (old but not obsolete saying)

During the Cold War, the whole system of arms and structure of armies and navies were created for use under conditions different from today, namely, for global war with the application of nuclear and conventional weapons. But now, and in the long term, the armed forces are intended for the solution of other problems, which have already been considered in previous chapters. In particular, strategic nuclear forces, as the most destabilizing component of the armed forces, are necessary only in the amount and composition sufficient for ensuring strategic deterrence. The role of air and naval forces has acquired new significance. For continental powers (for example Russia and China) the importance of ground forces will be preserved, although modified to fit new conditions and fulfill new objectives. Generally, among the composition of the armed forces, there is an increased need for a greater proportion of mobile forces, consisting of rapid reaction forces (assault forces with the necessary air and sea transportation means), and rapid deployment forces (forward units of ground forces, the air force and navy). Units intended for fulfillment of peacekeeping objectives and conduct of special operations are becoming more essential.

The structure of peacekeeping forces should be determined not only by the goals and needs of national defense, but also by the number of peacekeeping tasks entered on in the interests of international community. In view of the new requirements to the armed forces, methods of their manning, command and control, their organizational structure and direction of training of armies and navies will be improved.

1 PROFESSIONAL ARMIES OR CONSCRIPTION? (METHODS OF RECRUITMENT OF ARMED FORCES ON A MIXED COMBINED PRINCIPLE)

A significant influence on the character of armed struggle has always been exerted by 'the human factor' which in many respects depends on the method of recruitment of the army and navy. Much depended on the national homogeneity of standing armies. One could mention here, for instance, how much Russia profited from conscription which created an army firmly rooted in its country and tied to the interests of the homeland. This was very different from the mercenary armies of Western European states, manned by foreign soldiers.

Armies, both professional and recruited on the basis of conscription, have a long history of their own. In the Middle Ages, for example, princely and royal armies were the prototype of professional armies, and quite suited for the settlement of internal problems. But in the case of a threat of foreign invasion, it was necessary to recruit a volunteer corps. Later on, with the appearance of centralized states, the creation of national armies began.

Increasing numbers of wars using massive armies required the transition to general conscription in all large states, with comparatively reduced terms of military service for soldiers and junior commanders. Such a method of recruitment made for a closer connection of the army with the people, and the opportunity to accumulate mobilization resources trained in military arts, for deployment in case of war. In the majority of countries, such a system of recruitment has been used for about a century and continues today in various forms. After the war in Vietnam, the United States and some other states began the process of transforming their armies into professional ones, that is, they turned exclusively to contract recruitment of enlisted men and junior officers.

As P. Fouquet-Lapar wrote in his book *A History of the French Army*, throughout history the armed forces alternated between the two forms: an army of the people; and a professional army.[1] For example, England from the time of Cromwell adhered to the professional army, but during the First and Second World Wars was compelled to introduce general conscription. In America the desire to completely dissociate from the traditions of the Old World, predetermined the extremely negative attitude to conscription. In an official declaration of the US Congress in 1784, it was specified that in peacetime the regular army is not compatible with the principles of Republican government, and represents a threat to the rights of the citizens. But during the Civil War of 1861–65, and in subsequent wars, it has become clear even to the most ardent supporters of free choice that adhering only to the principle of free choice makes it impossible to ensure the strength of the armed forces necessary for wartime. As early as on the eve of the First World War, the American military leadership came to the conclusion that the principle of volunteer army had failed, and in 1917 Congress approved the Wilson administration bill on the introduction of conscription. In 1940, after France had been defeated, the law on selective service was reinstated. So it was with other states which adhered to the principle of volunteerism.

Russia and Germany, as a rule, adhered to general conscription. There was almost no large war that started and finished using professional armies. As a rule, the draft of additional contingents, and the mobilization deployment of the armed forces in view of the requirements of wartime, were required.

Therefore, when considering further prospects of the development of the armed forces, one cardinal question arises: whether to adhere to the principle of the creation of a professional army, or an army formed on the basis of general conscription. Of course, in speaking about the professional army one means first of all the level of its professional military training and structure, not just its way of recruitment.

It is obvious that an army consisting of volunteers devoting themselves to military duty, has greater opportunity for a higher level of combat readiness. But among the armies based on general conscription, a significant part (officers, employees and some junior commanders) is in essence professional. Hence, the key issue of the problem is whether to carry out recruitment exclusively by hiring, on a voluntary basis, or partially recruit the army by way of short-term service recruitment of soldiers and sergeants.

The authors from the Institute of Military History of the Ministry of Defense of the Russian Federation (V.V. Terekhov *et al.*), in the detailed review of foreign views on the professional army, notice that prior historical experience

> has no examples of the transition to a voluntary system of recruitment on the eve of war, but has examples of the rejection of this principle and the introduction of general conscription in periods of military danger. Though these examples add to the principal opportunity of such a transition, in practice it was implemented in countries (the USA, Great Britain), protected from the opponent by significant natural barriers.[2]

At the international colloquium 'Draft and the Professional Army' in Paris in 1991, and the symposium on this topic in Freiburg, Germany, in 1992, the peculiarities of various systems of recruitment of the armed forces were discussed. But the problem itself has predetermined the revealing of the advantages of a professional army.

Thus, to prove the advantages of a professional army, data from the Vietnam war experience were brought out, where the losses among draftees were larger than among volunteers (respectively, 57 and 34 per cent of the general number of losses). But the fact was not disclosed that the volunteers had the right to choose their assignments, avoiding the most dangerous assignments and specialties. Therefore, the draftees were exposed to a greater risk of being killed or wounded.

On comparing professional and recruited armies, many weak aspects of the latter were noticed, for the system of conscription has been repeatedly tested in numerous wars, its positive and negative sides are quite clear, and it is possible to evaluate them most authentically and on a well-grounded basis. As for the army recruited only by the voluntary principle, in the last 100–150 years no one country has had to solve serious military tasks by using such an army alone; it has not yet been tested in a big war. Therefore, the objective analysis and estimation, in view of all political, economic and military factors, is necessary without pulling out and over-emphasizing only the favorable or unprofitable aspects of this or that system. Probably, the model of military structure which would be faultless in all its aspects and could suit all countries in all times, does not and cannot exist. Various principles of recruitment do not exist in a 'pure' variant, they are mixed. Much will depend on the direction of military policy, economic opportunities, geopolitical situation, military traditions and specific tasks of the armed forces of the various countries at every particular stage of military construction. The variant that suits the USA could not completely suit Russia and vice versa. At the same time, the military practice and prospective requirements of military construction dictate some common, average requirements and principles which, in view of national specificity, should be taken into account to this or that extent in all armies.

In view of the above, it is possible to expect that in the future it will be more expedient for those states which can be threatened only by individual military conflicts, to have rather small professional armies, recruited on voluntary principles. For those states which will be compelled to consider the possibility of aggressive actions against them, or participation for those or other reasons in a large-scale war, it will be necessary to think of deployment of larger armed forces in wartime, and of preparing mobilization resources by training military contingents, systematically drafted into the armed forces, while maintaining a regular army in peacetime.

Generally, in spite of the variety of approaches, military construction in the majority of the countries of the world will probably be developed by a combination of voluntary, mixed and universal conscription.

Professional Armies, Recruited on the Voluntary Principle

The sophistication of arms and equipment, the manner in which they are used, and the conditions of fulfillment of combat missions predetermine for world military practice, the general tendency towards further professionalization of the armed forces, including those states where conscription exists.

Armies recruited exclusively on the voluntary principle are small in number (the USA, Great Britain, Japan, some countries of South America, and Africa). But in many respects they determine the shape of future armies, and certainly have many advantages. Their most significant advantage is the opportunity of selecting people who fit the requirements of military service, with regard to their education and state of health, and the great interest in military service on the part of the volunteers, who devote to it the substantial part of their life. The professional army with rather long service time for all categories of military personnel has an opportunity to constantly increase the level of combat training of personnel, to use modern arms more effectively, to have a high-level combat cohesion and coordination of units. It is not 'washed out' by frequent drafts and dismissals; the process of combat training proceeds with fewer interruptions and repetitions. A professional army provides fast transportation of forces to necessary regions, and immediate transition to active offensive actions, since conducting mobilization, especially in local wars, is not always required.

A professional army, recruited completely on the voluntary principle, costs of course a great deal, and not every state can afford it. But this aspect is usually exaggerated. Experience shows that great expenses are initially required only during the transition to such a system, especially for the creation of a high level of economic support of the personnel necessary for such an army.

In the USA, which had begun to shift to a professional army at the closing stage of the Vietnam war, the armed forces maintenance costs (in comparison with 1964) increased by US$22 billion, despite a general reduction of their personnel strength by 500,000, and the cost of maintenance of one serviceman increased 3.5–4 fold. Besides, the prevailing maintenance costs of the armed forces personnel greatly reduce opportunities for purchase of weapons, and investments in RDTE (research, development, testing and engineering). Only economically powerful states can maintain the appropriate level of expenditures in all these categories. In due course, the costs of personnel maintenance largely pay off thanks to reduced daily combat training, as well as to more skillful and careful operation of arms and equipment. A professional army has a number of other advantages which, on a whole, make it more organized and disciplined.

However, the necessity of the objective approach also requires taking into account the weak aspects of a professional army, recruited on the voluntary principle. They are generally reduced to the following.

First, the professional army, functioning as part of a closed system, has the propensity for isolation from society. The protection of the homeland ceases to be the duty of each citizen. The task of defending the country is assigned not to the whole society, not to the people, but only to a special military structure, which is inevitably connected with a certain withdrawal from democratic principles in the field of military construction. Such an army, though formally considered national, appears more dependent on ruling structures and can be easily used by them for political purposes. Financial incentives become of primary concern to those serving in the army. Thus, moral stimuli, connected with the fulfillment of military duty, are depreciated. Generally, the hired armies have never done very well in war. They can be quite good at carrying out tasks in short-term local wars against a weak opponent, but it is difficult for them to stand the test in a large-scale war, accompanied by great losses and long deprivations. When there is an appreciable risk of life, the material interest becomes secondary. In these conditions moral stimuli based on the unity of interests of the people and the army appear steadier.

Second, in a large-scale war, a professional army of peacetime proportions cannot act for more than for two or three weeks. Hereafter it must be constantly reinforced at the expense of the general mobilization of new forces. At the same time, being wholly hired under contract, it loses the opportunity to accumulate trained mobilization reserves, without which it is difficult to conduct a serious war. Even during the short-term war in the Persian Gulf region in 1991, the USA was compelled to call up more than 100,000 persons from the reserves. It makes one uneasy to imagine what could happen during larger military trials.

Certainly, when recruiting the army on the voluntary principle, a state system of the training of mobilization reserves can be created. For example, in the USA there is the system of the training of reserves through the structures of the National Guard and reserve component. But such training of reserves costs an excessive amount and does not meet all mobilization requirements, especially when one takes into account the rather rapid modernization of combat equipment, and the necessity of more frequent retraining of reservists.

Therefore, in the future, those states which do not exclude the possibility of participation in large-scale war, and where the recruitment of the armed forces is done only by contract, need to adopt laws to carry out short-term drafts on the basis of partial conscription, to create a more effective system of military training of reservists, and preparation of necessary mobilization reserves.

Third, in an army recruited by contract, where not only officers but also enlisted men have families, and generally live outside a military unit, it is difficult to ensure necessary daily combat readiness, even in peacetime. In an article by the Belgian General R. Close in *Spiegel* magazine (August 1976), and in an interview with the Supreme Allied Commander in Europe, General Alexander Haig, in response to this publication, it was said, and not without grounds, that if the armies of the Warsaw Treaty had suddenly assumed the offensive on Saturday or Sunday, when the majority of the military personnel of the NATO armies are usually at home, they could easily and in rather a short time move forward to the Rhine. The situation was relieved only by the fact that the countries of the Warsaw Treaty did not have such aggressive plans.

The complexity and contradictory nature of this dilemma is not always taken into account. Thus, the group of authors of the Russian Institute of Military History mentioned earlier express disagreement with the statements of some Western experts that an army of draftees is more adapted to conducting defensive actions, which thus testifies to the peaceful disposition of a country.

An army of professionals, on the other hand, can successfully conduct an offensive, which thus reveals the aggressive aims of a military policy.

> Following such logic, one could claim that in this case the lack of a particular skill is qualified as a guarantee of the security of the country. But then complete dissolution of the armed forces could become an even greater guarantee ... Any army can be referred to as a professional one only in the event that it is equally successful in any kind of combat in war and is capable to conduct both offense and defense.[3]

Here the essence of the problem is artificially simplified, and two completely different issues are confused – the combat readiness of this or that army for the beginning of war, and its ability to conduct various kinds of combat in wartime. The actions of armies depend on political objectives. Great Britain, during the 1930s, having a volunteer army, did not aspire towards war. On the other hand, Germany, adhering to conscription, conducted an aggressive policy.

As has already been emphasized, the professional volunteer army certainly has generally a higher level of combat readiness, and does not require mobilization especially when it possesses the initiative at the beginning of military actions and can proceed faster to the fulfillment of combat tasks. For the army which first needs to accomplish a general mobilization, it is more difficult to conceal preparations and assume the offensive. Irrespective of this or that intention, this is its objective peculiarity. While on the whole a professional army needs less time to be brought to combat readiness at certain times (for example, on days off, or at night), in the case of a sudden aggression by an opponent it might find itself lacking readiness to begin the fulfillment of combat tasks in comparison to the recruited army, where the majority of enlisted men and sergeants, as a rule, are permanently present at the place of location of units. If military volunteers were deprived of the opportunity to live in their apartments for a long time, they would just refuse to serve in such an army. Certainly, when there is no danger of a sudden aggression by the opponent for this or that army, this problem disappears. But even today, after the termination of intense counterposition of the armed forces of the two blocs, not all armies may consider themselves guaranteed from all accidents. Yet after the end of intense military confrontation between the two blocs, not all armies (say, for example, on Cuba, in the Middle East or Near East, in the Balkans, the Caucasus, or the American professional army in South Korea or the British Army in the Falklands) can consider themselves immune from any incidents.

In preparing for a transition to a professional army in Russia, other countries of CIS or Eastern Europe, it is necessary to take into account all aspects of the functioning of such an army, and the fact that it is impossible to carry out this transition at once. For this purpose, in view of political, socio-economic and other conditions and opportunities of this or that state, it is necessary to carefully think over the intermediate stages. For example, Holland, despite favorable conditions, plans to carry out a transition to a professional army only by 1998.

Among different states the transitional stages can vary, but under all circumstances the following are inevitable:

Due to the increase in the number of military personnel serving under contract (sergeants and enlisted men with important specializations), preliminary measures on making military service more attractive must be taken within the framework of existing military legislation or by amendments to it. Thus, in Russia it was planned that the number of career servicemen would by 1995 be raised up to 300,000 persons.

The propagation of service under contract, and the creation of favorable public opinion towards a transition to a professional army are required. The transition to contract service in those branches of the armed forces where mastering of more complex equipment (the air force, navy and rocket forces) is required should get first priority.

The second stage – the preparation and issue of new legislation acts on the transition to the new system of recruitment of the armed forces, passing budget appropriations for this purpose, increasing salaries, constructing quarters, granting a number of side benefits including vacations, study at civil educational institutions, medical service and social security during service, and employment upon retirement from service.

The third stage includes the completion of preparation and the passing of all necessary legal and administrative documents for final official approval, and creation of the mechanism for the functioning of the recruitment and military service based on the voluntary principle; correction of the whole system in light of developing conditions. Some legislative and administrative elements of military duty for a period of wartime, or in case of sharp decrease of inflow of volunteers for those or other reasons, can be preserved.

When seized by political zeal, certain members of parliament, journalists and some public figures in Russia and several other countries have demanded that the draft be abolished and the transition to a professional army begin immediately, they have not taken into account the difficulty of searching for optimal solutions and the ways of carrying them out which had been experienced by other countries.

In general, there is a big difference between the position of people preoccupied with theoretical, abstract discussion of issues, in full knowledge of the fact that they would not have to make any of the decisions, and the position of those who have to bear responsibility and solve concrete problems in practice.

The Bolsheviks in Russia before 1917 had said that the regular army is a reactionary force, and came out for the dissolution of such armies and the creation of national militia. On coming to power, they were compelled to immediately begin creation of the regular Red Army.

Lieutenant-Colonel Charles de Gaulle, in the 1930s, acted as a most ardent supporter of the professional army, but on being president of France for a long time after the war, decided not to realize his idea in practice. During his government, the principle of general conscription remained dominant in France.

Richard Nixon, in the election campaign of 1968, spoke out against the draft system, and in 1973 achieved its cancellation; but seven years later, after bearing the load of state responsibility, recognized that it was a major error.

Some Russian politicians and military men, who not so long ago were boldly calling for an immediate transition to the professional army, quickly cooled towards this idea, on coming to power.

The impatience of a certain section of the intelligentsia with regard to these issues is to be explained by more simple and practical reasons. They did not want their sons to serve in the army under any circumstances. As a result the Russian Army has once again become a workers' and peasants' army, just as it had been twenty years ago. In general, the transition to a professional army is a rather complex process, requiring much preparatory work. Under a hasty and irrational approach, with an insufficient account of the specific conditions of each country, the creation of such an army can only discredit itself. It is particularly difficult to carry all this out at the present time. For example, if in the USA it was necessary to increase official salaries three- or four-fold in order to raise interest in military service, then in Russia, taking into account the extremely meagre, almost symbolic money allowance of military men during basic military service, it would be necessary to increase salaries by an even greater degree. Nor does it mention the creation of social conditions in view of the general disorder in the Russian Army.

Another point is that for the creation of conditions for voluntary service, the appropriate transformations not only in the structure of the armed forces, but also of society as a whole, and especially development of market relations, are necessary. The experience of the USA and Great Britain has shown that the opportunity to recruit volunteers, depends on the ratio of the money allowance of military men to salaries in the civilian sector; on the system of privileges in society and the army and on the rate of unemployment. For example, it has come to light that a growth of unemployment by five per cent results in an increase of volunteers by 15–20 per cent.

Some Russian researchers, familiar with military service only from books and rumors, naïvely believe that as soon as there is an army recruited on the voluntary principle, such an army at once gets rid of the practice of using military personnel for agricultural labor, diverting them from combat training and other negative phenomena. But, if the economic relations in society do not change, the professional army will also continue to help gather the harvest, or solve other problems unusual for it, as it has been with the army based on general conscription. It is only possible to transit to a professional army together with the whole of society.

Recruitment of Armed Forces on the Combined Principle, Where Some Military Serve Under Contract, and Some Under Conscription

Such a system of recruitment of the armed forces exists at present in Germany, France, Italy, Spain and other countries. Within NATO states the 'contract part' of the military makes up on average 45 per cent. In Denmark and Belgium the percentage under contract is 60–64 per cent, in Germany 57 per cent, in Greece and Norway about 30 per cent, in Turkey up to 14 per cent. Even in one army, there are significant distinctions by branches of the armed forces. In France, the ground forces are manned with draftees by 55 per cent, the air force by 37 per cent and the navy by 26 per cent. The regular specifications in Germany define this ratio: ground forces 52 : 48, air force 72 : 28, navy 85 : 15.

If the intended plans are carried out, the contract proportion of military personnel in the Russian Army could be 30–35 per cent in the coming years.

On the one hand, with the combined system of recruitment of the armed forces, not only filling of the officer positions, but also other major specialties, connected to the mastering of complex equipment by the professional is provided for. While on the other hand, and to a greater extent than with recruitment only by volunteers – the public responsibility of the citizens for the protection of the homeland is retained, and more favorable conditions for the accumulation of mobilization resources trained in military arts, are created. Besides, smaller maintenance costs for the armed forces are required.

The experience of Germany is characteristic in this regard. During the planned transformations of its armed forces, it is preparing to reduce the personnel strength of its normal ready-for-action forces, along with the simultaneous perfection of the system of mobilization training. The activity of the future Bundeswehr is connected to mobilization reserves to a greater degree than at the present time. The realization of the new concept of training of mobilization resources began in 1995. It provides training for 140,000 reservists a year, at four training centers. Rank and file will be called for 24 days, sergeants for 45 days. The program 'Bundeswehr-2000' popularizes military service, aims at attracting volunteers, and prepares pre-conscription youth for military service on the basis of conscription.

The combined system of recruitment of the armed forces enables the use of certain positive aspects of the various systems of recruitment. At the same time, like any palliative system, it does not allow the realization of the advantages of any of these systems entirely. A non-uniform structure of military personnel and the legal division of them into two categories has many contradictions and difficulties, peculiar to conscription.

It is possible to consider a combined (mixed) system of recruitment of the armed forces intermediate between general conscription and the professional army. Despite the vague character of this system, its opportunities for further perfection have not been fully exhausted.

General Conscription

The third direction is connected to the continued improvements in the general conscription system. The positive and negative sides of this system are well known. Its main advantages are the all-national character of the defense of the country at lower economic costs, a more complete realization of the task of accumulation of trained mobilization resources, the interest taken in defense by the whole of society, and increased moral and political stimuli for the fulfillment of military duty aimed at defending the homeland. In this respect, the Soviet Army, for example, where within the basis of military service there was laid a conscious attitude towards the fulfillment of military duty, had relatively high morale and high combat potential.

Russia, as well as other republics of the former USSR, has now to determine the nature of their statehood and social systems, in order that the personnel of their respective armies may know what, and in the name of what, they are to defend. Without answering such questions, a French foreign legion or other form of purely hired army may exist, but a genuinely national conscript army cannot function viably in conditions of spiritual vacuum.

Along with the disappearance of the 'Soviet military threat', and the change of goals concerning external threats, as well as the general growth of pacifist trends in a number of countries, the problem of the moral and political justification of military service arises in NATO armies too. The already-mentioned General P. Foyard when considering the reasons for hostility towards conscription in some public circles ascertained: 'our consumer societies produce toxins of egocentrism, egoism, the cult of self, and deaden the spirit of defense. Military service is an antidote, it integrates society and carries out important social tasks.'[4]

There is a fundamental basis to the position of an army participating in the defense of the homeland, together with the people, or in any military conflict. When all the country is at war one is more prepared to risk one's life (here the behavior of a career serviceman and of a draftee are about identical). Entirely different though, is the soldier in a zone of remote military conflict, where the interests of his country are not so deeply-rooted. In the latter case it is more difficult to force the career serviceman to carry out combat tasks. For what sake has he to suffer severe deprivations, or to perish, when his own country lives in peace? It was not by accident that in the American Army in 1971 about 250,000 soldiers and seamen left their units without leave, and that more than 1,000 officers and sergeants were killed by their subordinates. As reported in the French press, with the direction of a small contingent of French forces to the region of the Persian Gulf, it became urgent to change draftees over to career servicemen, because out of every thousand soldiers it was possible to select no more than 12, which is why it was necessary to use combat units or to force draftees to sign a short-term contract.

However that may be, the career serviceman, as against the draftee, serves not in the name of duty established by law, but under a contract, which determines the rights and duties of the soldier, and allows its unilateral cancellation by the volunteer soldier. Certainly, the contract service does not exclude public enthusiasm, especially in periods when the country is in danger, but it cannot always be popular, for a career serviceman has other dominating stimuli.

The army that is based on conscription, is also more aware of the nature of war. Extremely limited opportunities in the choice of draftees, complexity of organization of military service and combat training because of periodic drafts of young reinforcements, and the discharge of military personnel who have served their term, are all basic deficiencies of an army recruited on the basis of conscription. After each draft, the unit is diluted by 20–30 per cent with young soldiers which thus reduces combat readiness for a period of 1.5–2 months.

Serious problems arise connected with maintaining military discipline. The national army recruited by draft, can only fully realize its advantages under conditions where general conscription is obligatory for all citizens, and is based on social justice. The whole history of the system of general conscription has been undermined, first of all, by infringements of social justice, such as the written or unwritten right of a part of the citizenry to buy themselves out of military service, to send someone else in their stead, to obtain the privileges of a delay from military service at the expense of others, and other evasions from the basic laws.

In addition, the situation in the armed forces, especially in the 'hazing' that goes on between various categories of military personnel, has reduced the prestige of military service and scared youth away from it.

Therefore, the acceptance of additional state measures on an increase of social justice, and the prestige of military service, with the improvement not only of morale, but also the financial incentives of military service, are important conditions for the further improvements and possibly the survival of the recruiting system of military service. The present constant reduction of terms of compulsory service is apparently possible only up to certain limits.

Within these questions great variety can be noticed. For example, in North Korea the service term is five years, in Sweden five months, in Portugal four months. The conditions of material maintenance are far from identical. In some countries, a soldier of basic military service gets more money than a career serviceman in other armies. In a number of countries (for example France and Spain) draftees are given the right to choose the place of service, time and type of draft, and specialization. Free air tickets for travel to places of residence and return to service on specific days, are given out. All military personnel are given the right to study by correspondence in civilian educational institutions, and those wishing to get a civilian profession during military service may enroll in special courses.

In Germany, Holland and the Scandinavian countries, military trade unions exist. Allowances are raised, and the volume of service duties is reduced. There are some who even say that if a soldier digs an entrenchment during a drill, he should be paid additionally. Distinctions in social positions of officers, NCOs and soldiers have been continuously reduced.

The listed measures on the democratization and facilitation of military service have, on the one hand, had a positive effect on the attitude of the military towards military service, but on the other hand, when they pass the limits allowable for military service, they inevitably result in the slackness of personnel, a downturn of discipline and self-discipline, and the primacy of some egoistic trends contradictory to the essence of military service which under all circumstances requires elements of self-restriction, selflessness and subordination. Otherwise the very reason for the existence of the army is lost. For example, here and there the practice of the inspection (the unannounced appearance of senior commanders in units on days off and holidays) has been abolished. It could, therefore, be possible to introduce any 'innovations' and restrictions on the activity of commanders and superiors, but the whole trouble is that the opponent may not take these restrictions into account and 'inspect' at a most inconvenient time.

As a whole, the humanization of military service is necessary, and improvements in mutual relations, in the moral and psychological climate in military units, and the increase of appeal of military service are especially important. But excessive attention to these questions can only be allowed in those countries which did not, and do not have, any dangers threatening them, and whose armies are necessary only as symbolic attributes of statehood.

Until recently, the possibility to increase the personnel strength of the armed forces were determined generally by the size of the male population. But in the last few years in the armies of various countries, the number of women has increased.

In Israel, conscription affects women as well as men. In the USA, a woman serving under contract can take any position, even be a Marine, but in the majority of countries in peacetime women are taken into military service only on a voluntary basis, and only in specified specialties.

During the first ten years of the existence of the professional army in the USA, the level of women recruits has increased by five times. The chairman of joint chiefs of staff G. Vessey declared in the mid-1970s: 'This is more important for our armed forces than the invention of nuclear weapons.'[5]

In almost all countries, a far-from-exhausted reserve stock of the professional part of the armed forces, at the expense of the increase of the quantity of civilian specialists, is currently in existence. In Germany for example, their number makes up more than 150,000 persons, while the personnel strength of the Bundeswehr is 495,000. At the same time, it is completely unjustified that the Russian Army contains plenty of officers within offices, in arms purchases, in financial, medical, housing and economic services, as well as teachers of humanitarian sciences in military training institutions and recreation facilities of military units, who can easily be replaced by civilian specialists. This would reduce both the proportion of officers among personnel, and the maintenance costs of the armed forces.

In Germany and some other countries, alongside purely military (with weapons) service, an alternative service exists, where the draftees work in the sphere of social maintenance, which is one of the reasons for the further reduction and abolition of conscription in these countries.

It is necessary to take into account the social role and influence of the army on society. As it exists in a number of countries, pre-conscription training of youth for service in the armed forces (developing physical stamina, and the acquisition of some technical specializations), are of considerable importance.

For the effective functioning of the system of military conscription the registration of pre-conscription youth by place of residence is required. Under the pretext of an infringement of human rights, the attempts to introduce such registration in a number of countries has caused counter-action from the public. In reality, the registration of the population in this or that form already exists anyway. If one ignores the legal side of the matter, taking into account only the administrative and organizational aspects of conscription, the experience of some countries proves that without organizing the registration of obligated reservists on a country-wide scale, it is practically impossible to carry out the due notification and draft of youth to military service.

While considering the problems of conscription, it is necessary to take into account some new issues which arise in the field of military training. Sometimes the matter is represented as if because of the growing complexity of weapons in the future, only a professional army would be able to master them. But one must keep in mind that arms, while becoming more complex in design, are at the same time equipped with means of automatic control and devices, which to a great extent simplify and facilitate the process of operation and control of combat equipment designed for shooting, driving, and fulfillment of other tasks. Therefore, new opportunities for mastering weapons in a shorter time appear, which make it inessential to have a professional soldier in each position.

The statement that general conscription is the destiny of only backward and economically weak states, is not valid either. The German historian Hans-Adolf Jakobsen, reflecting on German traditions of positive conservatism and the execution of radical changes in accordance with their objective development, writes:

> Historical experience and a number of good reasons speak for the preservation of general conscription in unified Germany. Till recently it was an essential guarantor of integration of the armed forces into

> society. General conscription symbolizes the belief that security is the common concern of all citizens and may not be simply delegated to some 'security agency'. In general, general conscription has become a safety lock from the revival of antidemocratic tendencies in the Army (from the danger of it becoming 'the state within the state'), (that has increased trust in the young German democracy in the country and abroad). General conscription must be understood as a legitimate child of democracy.[6]

To date, there are various forms of conscription in 83 countries. In some form or another, they will probably continue to exist for a long time to come.

In conclusion, it is possible to say that armies based on general conscription have not yet exhausted opportunities for the improvement of this system, including armies of economically powerful, advanced democratic states.

All this testifies that the 'correct' systems of recruitment of the armed forces do not and cannot exist. There exist only certain national interests, socio-political, economic, geostrategic conditions and potentials of states, and the objectives set before armed forces, all of which force a particular state to select the most efficient (appropriate) system of recruitment for its armed forces.

Of course, economic reasons have great importance too, but the main point is state expediency and military effectiveness, for, as the old saying goes: the defense of a country is equated to the price a nation is willing to pay for its sovereignty.

2 SOME QUESTIONS OF MILITARY CONTROL AND ORGANIZATIONAL STRUCTURE OF THE ARMED FORCES

On the Organization of Military–Political and Strategic Management of the Armed Forces

The processes of the democratization of society and the strengthening of the legal bases of statehood, will have an effect on the structure of the supreme bodies of military departments, especially from the point of view of integration of the armed forces into civil society. Two forms of organization of the supreme bodies of management of the armed forces have historically developed, and they will evolve according to the specificities of different countries.

The first one is connected to such forms of control as exist when the head of state is *ex officio* the commander-in-chief of the armed forces. With this system, a more precise and operational control of the armed forces is provided from the military point of view. Simultaneously, in the political respect, the required control over the armed forces is attained from the side of society which functions in a democratic and lawful way. Such control is, for instance, carried out by the Congress in the USA.

However, in countries where there is no developed civil society, the direct subordination of military departments to the president, as the supreme commander-in-chief, results in a situation where the army turns out to be in a special position, outside of civil and parliamentary control, and often all relations between the minister of defense and the head of state, or party, are built upon principles of personal loyalty, which is not only dangerous in the political sense, but harms the army, for, as much as any organization that is hidden from the light of day and the public eye, it begins to decay without noticing or covering up its flaws.

The second form of management of the armed forces, most completely expressing republican, democratic principles, is connected to direct subordination of the army to the executive power under parliamentary control. This is particularly so for Germany, where in 1956 the decision was taken to eliminate the rigid direct connection between the posts of the head of state and the commander-in-chief, by subordinating the armed forces to the executive branch of government. The adopted laws established a legal norm, according to which in peacetime, the power to give orders to the army is vested in the minister of defense – a civilian, and in wartime these powers pass to the chancellor – the head of the executive power. The army is under the control of parliament, the control functions of which in Germany are realized through the activities of the *Bundestag* defense committee, and of the representative of the *Bundestag* on military issues. Thus, alongside with military (army) administration, the presence of civil authority in the Bundeswehr is provided, which is not subordinated to military command.

In the opinion of German experts, such constitutional rules guarantee the primacy of political interests over military and serve as an effective barrier in the way of the possible evolution of the Bundeswehr towards a model of 'a state within a state'. Structures of military–political control similar to this one exist in some other western European countries.

The political advantages of such a system are obvious. But from the point of view of military efficiency it has some drawbacks, connected to the fact that sometimes military interests and reasons for the sake of which the whole structure of military management is created, are to a certain degree depreciated and removed into the background, and many military issues become complicated.

With any form of military control, the efficiency of its functioning depends in many respects on the clear distribution and efficient realization of the functions of civil and military structures inside the military department, and the absence of competition and duplication between them, with skillful supplementation of each other.

The worst version of the considered forms of military control is an artificial hybrid, when a civil person enters a military department as minister of defense, who then tries to not only usurp and solve directly military–political and administrative objectives, but also make decisions on operational-strategic issues, which are outside his scope of competence; as it was, for example, in the USSR with K.E. Voroshilov, N.A. Bulganin, D.F. Ustinov and other such leaders. In situations like this, no reasonable decisions can be expected on either of the above objectives or issues.

Of particular importance is the position of the general staff within the whole system of military command and control. We know that with the beginning of war, the commander-in-chief of the armed forces (president, chancellor or prime minister) takes upon himself the supreme military leadership, carrying out control over the armed forces through the general staff. The minister of defense (even if he is a professional military) becomes the deputy of the commander-in-chief. Therefore, the chief of the general staff, who in wartime bears the basic weight and responsibility of military control, remaining in peacetime subordinated to the minister of defense, should have the right to directly address the supreme commander-in-chief on strategic issues, and to prepare his work for wartime. Marshal B.M. Shaposhnikov in his famous book *Brain of the Army*, has convincingly shown that without maintaining this condition it is impossible to be prepared to successfully carry out military-strategic objectives. Recently, this problem was addressed by the chief of general staff of the Russian Army General-Colonel M.P. Kolesnikov.

If we adhere to the principle of 'teaching soldiers what is required for war', then this requirement generally, and not only during peacetime, should permeate the organization of military control at supreme levels. The effectiveness of command and control of the armed forces during war depends in many respects on flexibility and efficiency of political structures, and the skill to place the armed forces in politically favorable conditions. These, on the one hand, ensure the precise and realistic definition of military–political goals and objectives, the allocation of forces and means necessary for this purpose, supplies of material resources and constant state control over the fulfillment of set tasks; and on the other hand – the granting of relative independence to the military department in the selection of ways of fulfillment of strategic tasks, and in the organization and realization of operational control. Such a principle of organization of military–political and strategic management of the armed forces, was most consistently implemented in the democratic states.

In states under a totalitarian regime, where super-centralized and integrated political and military management of the armed forces during war was practised, such a system provided an opportunity for the maximum mobilization of all resources of the country in the interests of conducting warfare, and the direct influence of political authorities on the use of the armed forces. For example, for the Soviet Union, especially in view of the extreme conditions in which its war began in 1941, this circumstance had a certain positive meaning. But the adoption by both Stalin and Hitler of the functions of supreme commander-in-chief inflicted great damage on the control of troops. According to the science of control, it is impossible to supervise any matter successfully, without thorough knowledge of the essence of the matter. Incompetence, multiplied by excessive self-reliance, and the capriciousness of such dictators, including the inability to listen to and to understand the experts, as often as not simply turned into disaster. The point is not only the faulty decisions and actions. (They are possible with any form of government.) The situation becomes aggravated by the fact that the general staff and other bodies of the military department have their hands tied and are consequently unable to carry out their duties in full.

It is true that the division of political and military management makes for little change in a state with a super-centralized regime. In the beginning of the war [in 1941] the commander-in-chief of the Armed Forces of the USSR was People's Commissar of Defense S.K. Timoshenko, but he could not take any decisions without informing Stalin. Then during the war in Afghanistan from 1979 to 1988 the first deputies of the minister of defense and the chief of general staff remained almost constantly in position, but the plans even of rather small-scale operations were approved in Moscow. Nevertheless, even the formal holding of the post of commander-in-chief enables a military leader of firm disposition, such as G.K. Zhukov was, to achieve at least a certain degree of independence.

In general, the political leaders of western countries during the Second World War and in postwar years, were able, as a rule (except for Poland or France in 1939–40), to create favorable conditions for actions of the armed forces, and the all-round preparation of operations, while sometimes neglecting allied interests and proceeding exclusively from egoistic national interests. The USA entered the First World War only in April 1917, when the outcome of that war was already predetermined. They entered the Second World War only in December 1941, after the surprise attack by Japan at Pearl Harbor. The largest amphibious operations in Africa in 1942, in Sicily in 1943, and in Normandy in 1944 were conducted after a more than year-long preparation for each of them. This explains their relatively small losses. Combat operations in the Pacific zone had more serious consequences for the Americans. Military commands in the European and the Pacific theaters were largely independent. In this relation, the actions of the British government during the war with Argentina over the Falkland Islands (Malvinas) are also characteristic. The political leadership of the country had only to define the objectives, while the military command, without any haste, methodically prepared an amphibious operation.

Even today in the USA they frequently refer to the so-called 'Weinberg Doctrine', meaning that the USA can and should enter war only when conditions are favorable, and after a guaranteed all-round preparation for it. Rather instructive in this relation is the external political preparation of the USA for military operations in the region of the Persian Gulf in 1991: complicated diplomatic games, political intrigue with the countries of the Near East and with the President of Iraq, on instigating him towards aggression against Kuwait, all undertaken beforehand; preparation of public opinion and providing support inside the country and in the majority of other countries (including the USSR); conduct of well thought-out misinformation and psychological warfare with the purpose of exhausting and undermining the morale of the Iraqi forces; obtaining from the states supplying weapons to Iraq data on the frequencies of their radioelectronic systems; numerous delays to the date for launching the strike against Iraq, and the creation of the illusion of a possible political resolution of the conflict; the transition to the attacks by ground forces of the coalition after conducting a long and massive air campaign, and then only at the moment when Iraq had begun to withdraw its forces from Kuwait, and had removed them from prepared shelters and defensive positions out into the open. In short, everything was done to allow the troops to fulfill their assigned missions without serious risk, tension or great losses. This is the envy of any army, and any soldier.

But is it always possible? It is clear that this policy, this approach to the solution of military problems, could be undertaken only by such states as the USA or Great Britain; and also by other countries who were in an especially favorable geographical position during the Second World War when the main forces of the resisting opponent were held down on the Soviet–German front; or else in those specific political conditions which had developed by 1991.

For example Poland in 1939, France in May 1940, the USSR during the initial invasion in 1941 or in the battles of Moscow, Stalingrad or Kursk, had not the opportunity to choose either to wait until the completion of preparations, or to enter the war straightaway, to begin this or that battle or not. Once war has been imposed, battles have begun, and it is only possible to capitulate, or, irrespective of the degree of readiness, to conduct these unleashed battles. In such conditions completely different efforts are required from the armed forces, and the armed struggle acquires quite a different, more complicated nature with the infliction of losses.

From the point of view of exclusively favorable conditions for the conduct of war, the American army in the twentieth century was the luckiest army in the world – it had not been exposed to such severe tests as the armies of the Soviet Union, France or Germany. It is true that Vietnam was tough on the American army, but that was a specific war, a struggle against guerrilla forces, without large-scale battles of regular forces. But with the complex interrelations of contradictions in the multipolar world, and the chronic instability of conditions in a number of regions, the USA could be involved in one or another large-scale military opposition under conditions very different from the ones they would prefer. This is especially so now that recently the USA, without meeting any counterweights, has too hastily begun to get involved in various military conflicts, which are not always easy to withdraw from, or strictly regulate whether to participate in one or several of them. In the meantime, the USA remains militarily unopposed but too-expensive international political and economic actions, and frequent participation in military conflicts, could gradually drain and weaken the American state, as happened with the USSR as a result of the over-exertion of its economy in excessive military efforts.

Any system of organization of military–political and strategic management has its positive and negative sides. Their structure and functions in each war and in each country have their own peculiarities, and can be evaluated correctly only in view of particular conditions. But the tendencies towards greater complexity in the nature of armed struggle requiring the highest military competence and efficiency of control basically dictate the necessity of the future division of the functions of military–political, and purely military strategic guidance, giving the latter function to qualified military leadership. Under any circumstances, distribution of the functions on command and control of the armed forces, the structure and duties of control bodies, the points of control and the system of communication, should be determined and prepared in peacetime with respect to the requirements of war.

With the acceptance of defensive military doctrines, and on a pretext of the absence of distinct opponents, a number of countries have desisted from devising a full strategic planning of forthcoming actions for their armed forces. It is completely obvious that without a certain planning of the strategic application of the armed forces, it is impossible to carry out their purposeful construction and training.

The main thing in this planning is not the designation of directions of strikes (if any), but the reasonable estimation of potential threats, which can arise on these or those strategic directions (regions), and determination of the necessary force groupings, the sequence and order of their advance and regrouping, the accumulation of material resources, and operational logistic support.

Proceeding from this, the preparation of systems of communication and control, and of theaters of forthcoming actions, is usually carried out. All this indicates that despite arising complexities and uncertainties, the strategic planning of application of the armed forces remains a major task.

Prospects for Perfection of the Organizational Structure of Armed Forces

According to the allotted proportions of combat tasks, the role and ratio of the number of branches in the armed forces and other arms of the services, will vary. If during the Second World War the Soviet ground forces made up 80–87 per cent, the air force 6–9 per cent, air defense force 3–4 per cent, and the navy 4–7 per cent of an aggregate number of the armed forces, in afterwar years the personnel strength of the ground forces decreased to 30–35 per cent, and accordingly the strength of other branches of the armed forces has increased.

The development of weapons and military equipment as well as the combat capability of troops usually results in the occurrence of new combat laws. In the interval between the First and Second World Wars alone there emerged the air force, tank, anti-tank, chemical forces, forces of air defense and others. After the Second World War the strategic nuclear forces, space forces, units of radioelectronic warfare and antimissile defense appeared. Cavalry ceased to exist, and infantry and mechanized units transformed into combined arms and services. Some of the measures turned out to be not quite justified, for example, the allocation of the navy with the creation of independent naval departments in conditions where it was possible to solve strategic tasks only by the joint efforts of all branches of the armed forces.

The creation of strategic missile forces in the USSR was in general justified. The concentration of forces and means along with the development of air and naval nuclear components, allowed the USSR to reach strategic parity with the USA quite quickly. But the creation of the command of air defense, and thus organizational division of the armed forces of the country into five separate branches while promoting the development and introduction of new arms, has resulted in a kind of separatism of the branches of the armed forces, and has added a self-contained character to their activities. This has also complicated the interaction and control of troops, caused unjustified additional expenditures especially in the creation of bulky new control agencies, or systems of communication with command posts.

It would seem better in the future to organize the branches of the armed forces according to the natural spheres in which they act – on land, the sea, in the air, and in outer space. It is also essential to combine harmoniously various branches of the armed forces, and combat arms, according to their role and combat tasks, with unquestionable priority development of those branches which should carry out the main missions within the general framework of the armed struggle. But no branch of the armed forces may develop self-contained importance. The organizational structure of the armed forces can be strategically expedient and economically efficient only in the event of the various branches solving strategic tasks not by their own forces alone, but by supplementing and closely cooperating with each other. For an example, it might be useful to refer to the following episode from the experience of military construction in the USSR.

In 1978, during one of the command and staff military exercises on the preparation and conduct of strategic operations, the commander-in-chief of the navy, Admiral S.G. Gorshkov, formally and seemingly justifiably argued the need for the rapid expansion of Soviet naval forces, and for huge additional forces and means for the construction of submarines, anti-submarine forces, and aircraft carriers, in order to disrupt ocean communication lines between America and western Europe, and in particular, to prevent access by the USA and other NATO countries to oil-bearing regions of the Near East and the Persian Gulf. And, in fact, if one approaches the problem exclusively from the positions of the navy, one would think that there is no other way out.

At the same time, if one takes a broader look at ways of achieving the afore-mentioned strategic goal in view of opportunities of other branches of the armed forces, it becomes obvious that there are alternative, more effective, and much more economic ways of its fulfillment. For example, the creation of two or three tank armies in the Turkestan and Trans-Caucasian military districts in wartime, could provide the possibility (that is, if the war were to be unleashed by the opponent) to move quickly to the oil-bearing regions of the Near East, and to make supplies of petroleum to the NATO countries from these regions totally impossible. From the point of view of material inputs, this would be hundreds of thousands of times more cost-efficient, than the appropriate increase of naval power of the state.

Under modern conditions, huge, unjustified expenditures often result from the undying aspirations of individual services to carry out independently various missions in space; engage in electronic warfare; and, in particular, conduct the struggle against enemy aircraft and missiles exclusively by their own forces and means of air defense.

In view of these stated reasons, in due course it may be required to change the forms and content of the planning of construction of the armed forces in peacetime, and of combat control during war. They could be structured not by branches of the armed forces, but by strategic and operational-tactical tasks carried out by separate units of the armed forces (strategic nuclear forces, ICBM (intercontinental ballistic missile), submarines, strategic aviation), by strategic purposes forces (antimissile and air defense, anti-space means); and general assignment forces (the army, navy, front and operational-tactical aviation) including support forces in each component of the armed forces. In its turn, the necessity for transition to construction, training and combat use of the armed forces not by separate branches, but by compound components carrying out similar tasks, can require in the future the appropriate reorganization of structure of control agencies. At an intermediate stage, probably all states will adhere generally to a three-way system of the armed forces (the army, the air force and the navy) taking into account the objectively existing three spheres wherein they act. In some states the ground forces may not have separate command, being subordinated directly to the minister of defense and general staff.

The necessity of providing greater integration and effective coordination of actions of various branches of the armed forces, and cost efficient solutions to the problems of military construction can result in the requirement to transform joint chiefs of staff of the branches of the armed forces (in the states where they exist) to uniform (unified) general staff.

In those countries where air defense forces exist as a separate branch of the armed forces, it is probably necessary to transform this branch of the armed forces too. The fact is that in connection with the development of air attack means the dilemma arises: either to ensure the destruction of hundreds of carriers before they launch missiles, or to strike thousands of launched missiles which have smaller reflecting surfaces and a speed of MACH 5. Their destruction on the ground, in the sea, in the air and in space (at the start of their launch) demands the enlistment of all branches of the armed forces, and first of all, strike means.

In addition, as already emphasized, military actions will in the future start with and be accompanied by prolonged air operations, and even campaigns. Without their successful conduct and reliable suppression of the enemy by the NATO command, no offensive by ground forces can even be imagined. This means that the successful repulse of massed air strikes of the opponent and the defeat of his aviation can largely predetermine the failure of his offensive operation. Therefore, rapid actions from all branches of the armed forces on overthrowing the air groupings of the opponent, along with air defense and space defense, gain the decisive emphasis in modern armed struggle.

The task of planning a strategic operation to repulse an aerospace aggression of the opponent, the organization of control and interaction of all forces, including those involved in active struggle with the air opponent, could be entirely solved just by the general staff. Space systems and missile early-warning units should be subordinated to it, as they work in the interests of all branches of the armed forces. It is necessary to drop the existing system of air defense and to create a unified system of air defense of the country and the armed forces. The character of future operations also requires a significant increase of the role of command of military districts in the struggle with air opponents. Therefore, it is expedient to have radio-technical, air and missile defense forces under the structure of combined arms services and fighter aviation – in the structure of air units. Russia and countries of the CIS need to develop strategy for the creation of an effective air defense system. NATO faces the same problem.

Peculiarities for Organization of Control on the Operational-Tactical Level

The agencies for control on operational-strategic and tactical level are also very significant. During the First and Second World Wars, command and control of force groupings on major strategic directions was provided through agencies of control for the fronts (groups of armies), which included some combined arms, tank armies, air armies and other units of combat arms. The necessity for combining efforts of several fronts required the preparation and conduct of operations of groups of fronts under the direct management of general headquarters (*STAVKA*) and the coordination of their efforts by *STAVKA* representatives.

Considering the nature of future operations, it is thought most expedient to create commands without division by fronts and groups of armies, for the control of diverse force groupings. For some states (the USA, for example) it might be acceptable. But for such continental countries as Russia, China or India, with their extensive territories and possible simultaneous conflicts with different opponents, it would be impossible, especially in large-scale war, to do without the creation of fronts.

There are a number of new problems in the organization and control of forces on the operational-tactical level. Taking into account that in the armed struggle of the future, combat will develop non-uniformly, some force groupings developing a prompt offensive may pull out ahead, while others may cover flanks or repel counterstrikes, fight in encirclement, or conduct other types of combat. In these conditions, units should show the greatest possible degree of independence and initiative in the execution of combat tasks, as well as high maneuverability. In view of all this, their organizational structure will be subsequently developed. The ground forces units will become less bulky and even more mobile. At the same time, their firepower and strike force will increase.

The variety of theaters of operations, the organization and technical equipment of the probable opponents, and the character of tasks to be executed will still result in the need for infantry (mechanized infantry), and mechanized, and tank units. For large-scale war, continental powers may require not only combined arms and services, but also tank corps and armies to be used as powerful strike forces on decisive directions.

In the mid-1970s in the Soviet army the hypothesis was put forward, that in connection with the high mobility of all units and combat arms, there was a need for tank armies as well as for the creation of mobile groups, which were intended for the exploitation of the success of an offensive on main directions. But two important circumstances were not taken into account. First, even with the high mobility of forces in general, in the interests of the prompt development of an offensive into the depth and to the flanks, there is always a requirement for even more mobile units, and even better equipment. It is impossible to equip all troops at once to an identical level with the newest combat material. It should first of all be supplied to the units specified to solve such tasks.

Second, the faster progress of mobile groups or other similar groupings in offensive operations is reached not only at the expense of equipping them with more sophisticated equipment, but also by creating more favorable conditions for prompt actions by means of massed employment on their axis of the major effort of aviation, artillery, airborne and engineering forces, and the use of other means of reinforcement.

Subsequently, therefore, the thesis on the uselessness of special more-mobile force groupings in the Soviet army was reconsidered. It was decided to create operational maneuvering forces instead of mobile groups, as used during the Second World War. Their main difference from former mobile groups was that not only tank armies and divisions were used, but also separate army corps of separate structure were specially created for actions as operational maneuvering forces, while within the main body of the army tanks, mechanized infantry, artillery and other units were incorporated, equipped with the latest T-80 tanks, amphibious infantry combat vehicles and armored personnel carriers, and self-propelled artillery. For the first time, the structure of these army corps included an airborne assault regiment and army aviation. If the American army and other NATO armies believed that any forces are capable of conducting an air–land operation, in the Soviet army special forces specifically intended for actions on the spearhead of these operations were created. Together with the Soviet Armed Forces, such army corps which were meant to act as operational maneuvering forces, have now been liquidated. But, as it has already been remarked, it is impossible to liquidate operational-tactical, as well as technical ideas, and consequently with further perfection of the organizational structure of troops, they will probably be used in one form or another.

Which organizational structure of combined arms and services units (infantry, tank, mechanized) can be subsequently realized? In this relation two tendencies are seen among the leading states of the world. In the USA, Germany, France, China and some other states the division structure is retained. In Russia and some countries of the CIS there is a move away from armies and divisions, and talk of the expediency of a transition to army corps consisting of several brigades. Previous experience, and the prospects of development for ways of conducting military actions, suggest that the corps and brigade structures appear expedient only for mobile forces. It is better to have main force groupings intended to act as security echelons, and for subsequent escalation of efforts, in a division structure.

The divisions of the US Army consist of 10 to 12 separate battalions, which are united by brigade command only temporarily, during the execution of certain tactical tasks. In the Soviet, Chinese and other armies, divisions traditionally consist of regiments, including certain units on a permanent basis. The strictly theoretical and still untested superiority of the American organization of the division lies in flexibility of divisions in tactical relations, which allows when necessary a brigade of four to five battalions on the main direction, and two to three battalions on minor directions and so on. But when conducting intense combat all this will look much more difficult. At the next regrouping of the division (and this can occur often even within a few days) it is necessary to organize brigade groups anew, and each time they appear insufficiently cohesive for joint actions. It is necessary to redistribute reinforcements to organize interaction and control of units. The logistic support bodies cannot always be reconstructed in due time, in view of the new reorganization of battalions and artillery, and other units for the supply of ammunition, fuel and other materials, or the removal and medical treatment of the wounded. In difficult battle conditions, organizational and other disorders with inconsistency of actions are possible, producing a negative effect on the course of combat.

Long-term battle experience shows that there need to be constant and steady army groupings at the tactical level – first of all, regiments and battalions with regular artillery, and anti-air and engineer means; in which units that work in common and become used to each other reach a high degree of coordination and cooperation of actions. In conjunction with this, the massing of means on this or that direction, and the transfer of efforts from one area to another by divisions, may be achieved by increasing or reducing the zones of action of regiments, by the massive concentration of means, and by switching the main forces of the division artillery and aviation to the necessary directions.

The search for the optimumal parity of weapons and personnel strength remains one of the important problems of organizational construction of the armed forces. It was originally supposed that the creation of automated systems of weapons and troops control would allow the reduction of the teams and crews serving such weapons and equipment as well as the personnel strength of signal units and control agencies. But in reality, it turned out that such a reduction of personnel strength does not necessarily occur in all armies. On the contrary, an additional number of people are required to maintain automated control systems, and the number of control agencies and supporting units in general, grows even more. This problem awaits solution.

Combat maneuverability will probably require the dismantling of bulky divisions with a strength of more than 15,000–16,000. But more sophisticated means of control make possible an increase of weapons and equipment in units: in tank and infantry companies and artillery batteries of up to 15–18 tanks and APCs (motorized infantry combat vehicles) and eight guns, up from ten tanks and APCs and six guns.

In the second half of the 1980s, in connection with proclaiming their defensive military doctrine among the countries of the Warsaw Treaty, offers were put forward that units should generally have only anti-tank, air defense, and other specifically defensive types of weapons, and that tanks and other offensive means would be excluded (or reduced) from their structure. To give the forces an exclusively defensive nature it was even proposed to disband the majority of pontoons and ferrying equipment, not taking into account that in a defense lacking forces and means, maneuvering them from one direction to another is of especially great importance, so that in defense ferrying equipment is no less necessary than in an offensive. However, giving divisions only an offensive or defensive character is extremely irrational from all points of view, because troops should be ready to solve all combat tasks.

One of the new peculiarities of military structures is the creation of multinational army corps and divisions. Previously, the principle of multinationality in NATO's armed forces was spread only with reference to formations up to an operative-strategic level – group of armies, ATAF – it is now proposed to expand to the level of army corps or divisions. The rationale behind the creation of such formations is probably political: to strengthen the military integration of the NATO member states, to increase mutual trust and mutual control (especially concerning the German armed forces), and to ensure the continued stationing of American forces in Europe.

Such a form of integration could probably be useful for the countries of the CIS. In an arrangement of September 1993 between the ministers of defense of the USA and Russia, the grounds for the subsequent creation of a peacekeeping army corps, including a division of the Volga military district, and a division of the US European command, were in essence achieved. The USA also has certain arrangements on this account with the military command of the Ukraine and Belarus. International forces of the CIS for actions in the region of the Tajik–Afghan border have also been created. Such forces will probably arise in the Caucasus too. Therefore, such forms of organization of international forces are now quite likely.

The dynamism of combat requires the appropriate maneuverability and mobility of the rear logistic support bodies and units. Therefore, the creation of an advanced infrastructure of logistic support in theaters of military action, designed for the supply of mobile forces which can be moved to these directions, in addition to central logistic bases, seems quite in order. But the attempt to assign all the basic tasks of the material maintenance of troops, navy forces and aviation forces only to these mainly stationary logistic bases, by reducing logistic bodies and reserves in units and formations, is an extreme move.

The transient and maneuverable nature of contemporary combat, the increasing needs of forces for ammunition and fuel, the losses in personnel and battle equipment, the opportunity for the destruction by the opponent of logistic bases located in the deep rear give great importance to the independence of units on questions of material maintenance, and their ability to carry out the repair of arms and equipment on the battlefield. It is necessary for units to have a minimum necessary amount of material stock (regiment and brigades for 3–4 days, divisions for 5–6 days) and of logistic bodies, capable of solving problems of logistic support independently and more effectively, within the limits of norms of troop stocks. This is also of great importance for the increased survival of the rear as whole.

If in this relation one compares the American and Soviet formations and units, there has always been a large difference in the ratio of strength of the combat and support units, especially of repair ones. The proportion of support units in the American army, as a rule, was 2–3 times more than in the Soviet army. Many critical articles were written about these bourgeois overindulgences. But the aspiration of the Soviet staff bodies to 'save' on maintenance units in peacetime and in wartime usually came to an end by the distraction of combat units for the fulfillment of economic and construction tasks. Probably, in considering the prospective structure of forces, it is necessary to search for the 'golden mean' of the rational combination of combat and support units, in the structure of formations and units.

In the future, operations and combat will be characterized by increased scope, the participation of various branches of the armed forces and combat arms, equipped with diverse, complex, military equipment. These operations will possess high dynamism and combat maneuverability and will be conducted in conditions of the absence of continuous fronts with sharp and fast changes of conditions. Both sides will be engaged in a fierce struggle to seize and maintain the initiative and will face strong radio-electronic counter-measures. They will also face far more difficult challenges to their command and control.

With the greater speeds of missiles and aviation and the increased mobility of troops, especially in the system of strategic nuclear forces, air defense, and air forces, the combat control function will more and more take on the nature of executing pre-planned variants of decisions, of programming, and of planning for anticipated combat actions. The high level of planning of operations and combat actions will become the chief precondition for successful troop control.

The automation and computerization of control requires improvements not only in the organizational structures of control agencies, but also in forms and methods of work of commands and headquarters. The latest achievements of science testify that the control system can be effective only if it develops both along the vertical and horizontal axes. It means, in particular, where the principle of unity of command is held, a considerable extension of the front of work, granting more rights to the chiefs of combat arms and services. These chiefs should solve many questions independently, coordinating them with combined arms and services headquarters and among themselves, as with extremely limited time and rapid development of events, a commander cannot consider and personally decide all major questions for the preparation and conduct of an operation, as in the old days. Considerably greater initiative and independence on all levels is necessary.

It is also necessary to reevaluate the forms and contents of combat documents, so that they are less bulky and more specific. Otherwise, their computerization becomes extremely difficult. Experiences from combat exercises show that it is impossible to integrally join modern automated control systems with the old documents and methods of work.

A great amount of work on the complex solving of problems of automation, and perfection of traditional methods of control with the aim of their approximation to principles of computer systems has been done in the troops of the Far East, where General K.I. Kobets acted as a generator of basic ideas in this area. This effort demands further development.

The perfection of the organizational structure and control of the armed forces will also be proceeding in many other directions too.

Thus, future prospects for the construction of the armed forces consists in equipping them with new types of precision weapons, in the perfection of the organizational structures and the system of recruitment, and the organization of control.

NOTES

1. P. Fouquet-Lapar, *Histoire de l'armée française* (1986) pp. 3–4.
2. V.V. Terekhov *et al.*, *Professional'naya armiya za rubezhom. Istoriya, opyt i usloviya sozdaniya* (Moscow: Institute of Military History of the Russian Federation, 1993) pp. 158–9.
3. Ibid., p. 165.
4. *Revue de défense nationale*, No. 1, 1990, pp. 35, 37.
5. Cited from: S.V. Yermachenkov, G.M. Ivanitskiy, V.V. Terekhov, *Zhenshchiny i voyennosluzhaschie v inostrannykh armiyakh* (Moscow, 1991) p. 35.
6. Kh. A. Yakobsen, *Ot strategii oborony k politike mira. Germanskie vooruzhennye sily v periodu peremen* (Moscow: Nauka, 1993) p. 138.

CHAPTER 5

BASIC DIRECTIONS FOR PERFECTION OF TRAINING OF THE ARMED FORCES

As historical experience shows, the combat effectiveness of an army, and successful warfare depends on many factors: the political goals and economic capabilities of the state, the personnel strength and armaments of forces, their training and combat morale, military art, the abilities and resolution of actions of military leaders, and other specific conditions. Among the last mentioned, one of the most important conditions determining victories and defeats has always been the level of training of commanders, staff and troops. The problem of training of the armed forces is connected with the resolution of a two-fold problem. First, it is important to envisage correctly the possible nature and ways of conducting future war, and accordingly determine the components of military training. Second, it is necessary to define training methods for troops, to find such forms and methods of training which would provide the realization of advanced approaches to ways of conducting war, and corresponding requirements for the training of forces.

Both sides of the problem are interconnected, and require a systematic approach and combined solutions of many complex tasks of development and achievement in military art, and of theory and practice in military training and education. A reasonable determination of the contents of training, and the development and introduction of everything new in the process of combat and operational training, in accordance with the reasons stated in the previous chapters, has especially great importance. Training and conducting armed struggle becomes more and more complicated and sets greater requirements on the training of commanders and staffs of ground, air and naval forces.

The army, conscious of its professional responsibility, should use this source as much as possible. Further perfection and increase in effectiveness of military training and education of officers and all personnel is one of the major and inexhaustible sources of enhancement of combat powers of the armed forces, practically without any additional material inputs.

To be ready to accomplish any task which may arise in combat conditions, headquarters, troops and naval forces should be trained for all kinds of combat. After the Second World War, the armies of NATO and the Warsaw Pact simulated mainly offensive operations and combats. Recently, in the west, materials and maps of the German Democratic Republic general staff were published, indicating plans for offensive operations. But it was not mentioned that in all exercises where such plans were used, the operational conditions were simulated to allow the armed forces of the Warsaw Treaty Organization to switch to the offensive in response to aggression on the part of the opponent. I happened to be the developer of many such exercises and I know well that the forces of the Warsaw Treaty Organization were not permitted to initiate military actions. But offensive operations, before intrusion on to their territory by countries of the Warsaw Treaty Organization, were practised by NATO armies too. Information on similar NATO exercises has not yet been published, though it would be fair to make public the plans of the exercises of both military blocs. But even so, some of the exercises of the NATO forces are well known, and where offensive actions were worked out.

In this book the schemes of some exercises are given. It is quite clear that in modern conditions, armies that adhere to the defensive military doctrine cannot practise exercises in which unleashing of aggression or the initiation of any military actions, are envisaged.

The manifest openness in military art now reveals new opportunities for joint conduct of exercises by the armies of different countries, and an exchange of experience of combat training.

This should be favored by understanding that the artificial division of military science into 'socialist' and 'bourgeois' contradicts the essence of this science, because the laws and the principles of armed struggle are objectively uniform for both sides participating in counteraction during war. Certainly, the armed forces of each country have national specificity. For example, for some countries aircraft carriers are developed in military theory and practice to different degrees. It is possible to speak about American, Russian or German military science only provisionally, meaning the contribution of each one to common knowledge. As a whole, while military science absorbs knowledge and experience accumulated in different countries, it is uniform for all countries, like any genuine science.

The correct guidance of officers, NCOs (sergeants) and soldiers on the nature of armed struggle, including nuclear war, has an exceptionally important meaning. Not long ago a paradoxical situation had developed. It was becoming more and more obvious that nuclear war is an impossibility and that it is senseless to adjust force structure and training exclusively to such a war. It was necessary to reconsider not only operational-strategic and military– technical, but also ideological concepts as well. At the same time, critical statements concerning the under-estimation of the influence of nuclear weapons on military art and of revolutionary transformation in military affairs were heard most often from political and military men, writers and publicists, who most zealously opposed the concepts of nuclear war.

In spite of the fact that the concept of nuclear war was itself disposed of, and that it was necessary to prepare personnel of the armed forces for protection of the homeland under a completely different nature of armed struggle (without application of nuclear weapons), the ideologists continued to demand the rejection of military-patriotic education of pre-conscription youth and military personnel, and instead the dissemination of their 'anti-war-patriotic' education. A noted writer, A. Adamovich, even offered to create a special 'super-literature', which could expand anti-war consciousness. He wrote 'may a bomb blow up within ourselves, and in the consciousness of literature.'[1] But why on earth educate an army into a contempt and hatred for the profession, and for military duty?

Activity directed towards the eradication of war and the activation of public peace movements deserves support, including ones coming from the military, and war veterans. But this should not confuse people who prevent wars and constrain aggression, if necessary by military means, in the interests of society. In any army, servicemen should receive correct and precise orientation on questions touching on the essence of military service.

Problems of Officer Training, and Regulation of the Military Way of Life

The improvement in the quality of the officer corps, as the key, basic link in the chain of complex problems connected with new trends in the training of the armed forces, is the most important condition of the combat effectiveness of the army and navy.

In different countries this problem is solved in various ways. But when aiming at the preservation of the stability of its officers, any army should first of all determine the number of officers and specialties to be kept in the next 10–15 years, according to the reasonable forecast of prospects of change in the organizational structure, personnel strength of the armed forces, and their technical equipment (the types of weapons and equipment that will be used by the armed forces and the navy). In view of this, the numbers and structures of military training institutions, and their terms of training, could be determined.

It is also necessary to carry out thorough professional sociological research to find out what a modern officer should be, and what requirements he should meet in view of the varied conditions of military service, tasks, and nature of armed struggle.

The main thing in the training of officers at the present stage, especially in the long term, is a high level of general culture, moral qualities and military professionalism, that first of all assumes the transition from a superficial study of military theory to the development of a creative foundation in mastering military art, and also organizational qualities, which are necessary to exercise command and control of troops in combat, and the direction of their training and education in peacetime.

This should constitute the basis for development of realistic programs of training in institutions of higher education for commander training with troops, and the extension of officer training courses. The programs should be correlated in order to constantly increase the level of training of officers in accordance with their advancement in service. It also seems expedient to take into account new requirements for military personnel in the appropriate regulations of service of the officer corps, in combat and general basic military regulations and other documents, regulating military service.

Thus, the main point in the transformation of systems of training of personnel, as well as in other sections of military reform, must not be the specification of general requirements, but the development of such organizational measures and mechanisms, which would stimulate and guarantee the realization of new requirements regarding the training of officers and the system of military service. As we know, in Russia and other countries of the CIS measures on the perfection of systems of military training have now been taken. But the majority of them are generally reduced to organizational questions, and to changes in terms of training in schools, academies and military colleges.

Certainly, in many respects these measures are justified. A reduction in the number of military schools and academies with their integration has long been awaited. But along with this it is necessary to create a modern educational base in schools and academies and to provide more efficient maintenance units in order not to distract cadets from study. In some armies, the development of educational programs involves a struggle for each academic hour allocated for studying any subject, but in reality the curriculum is used irrationally. Sometimes up to 20-25 per cent of educational time (academic year) is spent on economic labor service duties, and routine maintenance.

The preservation of army political schools seems to be unjustified. In the majority of armies in the world nobody prepares specialists for political and educational work separately. The army needs officers specializing in issues of the education of personnel, and also psychologists and sociologists. But future officers should first get a military speciality, and then, after graduating from command or engineering school, undergo retraining in the specifics of educational work. It is impossible to carry out moral and psychological training and education of personnel when out of touch with the specificity of combat activities of combat arms and services.

Long-term experience also shows that any organizational measures will not give proper results, if pedagogical processes are not transformed fundamentally. Therefore, in view of modern requirements, it is necessary first of all to reconsider the contents and methods of the training and education of officers in military training institutions. Along with an increase of professional standards, it is desirable to pay special attention to the development of pedagogical skills in an officer. First of all, it is required that a commander possesses the ability to convince people. For this purpose an officer needs to possess thorough pedagogical and psychological knowledge and skills. Therefore, it is desirable to introduce a pedagogical course in all military schools. Working with people has become more complicated with democratization and the appearance of other new phenomena in society. A particular problem is inculcation of the sense of dignity and honor in the officer. For this purpose the system of training and education of cadets and students should be totally changed. It is necessary to teach them and bring them on not collectively, but to put the main emphasis on painstaking individual work on the molding of personality and the forming of each officer.

In this relation there is much that is instructive in American military training institutions. The educational process there is characterized by great freedom of creativity and personal initiative, proceeding from the principle 'teach not what to think, but how to think'. The main emphasis is not on the 'coaching' of officers in certain questions and specifications, but on an independent acquisition of knowledge and development of thinking.

Quite democratic conditions at the lectures, spontaneity and liveliness of dialogue among the teachers and students are eye-openers. Computer equipment and various modern simulators are widely used during training. The authorities of the US defense department and of the branches of the armed forces believe that the main lever of influence on the state of affairs in the armed forces, is the work done in higher educational institutions, along with organization of the training of officers. In particular, any sample of military equipment being adopted is always accompanied by the appropriate simulators, which are first directed to educational institutions and centers.

American officers, in comparison with Russian officers, imbibe a broader fundamental general education and culture. However, the graduates of Russian and German military schools are superior with respect to tactics, the practical skills of small-unit command, shooting, and the operation of equipment. It is enough to say, for example, that during four years of training, cadets at West Point shoot only once per basic kind of weapon, and practice tactics only once a year, during one to one-and-a-half months at summer training centers. On graduation from the school they are not yet high-grade officers, and fill positions only after additional training in a speciality at one of the training centers.

With further perfection of the educational process in military training institutions, other armies probably should not copy this or that system, but take the way of extending and deepening general educational training, in combination with development of the necessary practical skills.

The most difficult and complex problem in the path to the reorganization of military training institutions is the selection and education of the teachers. It is necessary to retrain teachers so that they can teach using new methods. In some armies there is still a persistent custom to direct to military training institutions those generals, admirals and officers who display themselves poorly at command and staff work; and on being assigned to military training institutions they practically lose any prospect of advancement in service.

It is necessary to stimulate the work done in military training institutions, and to ensure a combination of service both in military training institutions and also among troops, and to promote the best generals and officers to schools and academies. For example, in the USA after the Persian Gulf War, a significant part of the officers who distinguished themselves in battle, and obtained combat experience, were directed to work in the National Defense University, in military colleges and training centers at Forts Leavenworth, Knox, Benning and others.

The stimulation of the military service of officers has exclusively great importance from the point of view of validity and fairness of promotion, and in awarding military ranks. In this relation, the plain succession of promotion, and awarding rank by experience and principle of seniority, in precisely established terms, irrespective of attitudes to service and abilities, similar to that in the British army and some other armies, is not quite justified.

At the same time the system of the Soviet army, where minimum compulsory terms of service for an officer in this or that position were not established, also proved insufficiently effective. Such a system creates in essence potential opportunities for the accelerated progress of the most capable officers, but it also lays grounds for subjectivism and arbitrariness, which displeases many officers.

In view of the pros and cons of various systems, it seems appropriate to determine a minimum necessary term of service for officers in the basic command positions (not less than two or three years), as is practised in a number of armies. The absence of such minimum terms allows some generals, admirals and officers to serve at excessively accelerated rates, complicating the acquisition of the necessary experience, and dooms others to an extensive standstill in the same positions. Certainly, there should not be any wage-leveling. The early promotion of the most capable officers is appropriate, but within reasonable limits.

It is desirable to direct main efforts towards an increase of authority for conscientious and competent officers, taking into account the level of their training in awarding the next rank and promotion. In this connection, the practice of the evaluation of all officers every four to five years, existing in some armies, does not justify itself. In general and simultaneous evaluation, practically all officers would be reviewed without any need, even those who have been in a position for one year. Such an evaluation results in haste, a reduction of quality, and an inconclusive review of officer qualifications. The complete and simultaneous review of all officers does not meet the requirements of strictly individual approach to each officer.

A permanent evaluation system assesses each officer after he has served the established service time in a given position, and in accordance with his preparedness to be advanced to the next position, or be directed to training. This system can exist in each formation and unit, and the attestation should be preceded by special examinations or retraining.

It is necessary that each officer understands and is convinced that successful promotion depends first of all upon himself. It is desirable to stimulate the interest of an officer to the constant increase of his professional competence, so that not only senior commanders, but first of all each officer himself is concerned for his training. It is impossible to get things moving without overcoming the passivity and indifference of officers.

Additional moral and material stimuli capable of getting officers interested in zealous service are necessary. The best appeals, the best ideas for the improvement of a situation of an officer will remain mere talk if they are not combined with personal interest. In particular, it is necessary to reject the idea of payment of additional salary at the end of the year, to practically all officers. These means are needed to give an incentive to officers for real practical results in their service.

Among the CIS countries there are political and military figures who still propose that while society as a whole shifts to market practices, the army should be built exclusively on socialist principles, as in the past. Without question, conscientiousness and service as a duty will always be necessary. But market practices demand that the soldier and officer be appropriately compensated during military service and that their living standard be on a par with the middle strata of the population.

We know that in the armies of socialist countries, along with commanders, there were military commissars, subsequently political assistants. This measure was partially justified only in the Red Army during the Civil

War, when many officers of the Tsarist army served in the Red Army, so that commissars were necessary to supervise them. But in subsequent years, and especially during the Great Patriotic war, the institute of military commissars has not justified itself. Life shows that any military system can effectively function only on the principle of unity of command, carried out on a legal basis.

Apart from the commissars, there are other forms of constraint over the activity of commanders and chiefs, especially when the activity of officers is regulated in detail, by basic military regulations, manuals, guides and orders.

The military press has repeatedly published proposals on an increase of independence for commanders in directing combat training, strengthening discipline, and in economic activities. But the majority of these have not been realized. Therefore, the resolute rejection of unjustified centralization, for the petty regulation of officers' activities is required. Excessive over-sight and the excessively great number of inspections by the almost continuous presence of senior chiefs in units create tension and have a negative effect on the mood of officers. It is desirable to give commanders real independence and initiative in planning and directing combat training and administrative activities. But all this cannot be changed by general demands and appeals. The new situations and approaches should become the basis of new basic military regulations, and other basic orders and instructions, which determine functional duties, the amount of inspections and certain other measures.

As a whole, the entire system of measures should make the service of officers more organized and confident, mobilizing and stimulating a zealous attitude to the fulfillment of military duty. All these questions cannot be solved in isolation from each other. Only composite measures, coordinated with the entire system of military service, including the increased role of junior commanders and personal responsibility for each soldier, will enable a cardinal change in the qualitative structure of the officer corps. This is particularly important for armies that retain general conscription. One of the important problems is upholding the role of the junior commander in the training and education of subordinates.

In the Soviet army, and now in the armies of the CIS, the role of the NCOs – sergeant and master sergeant – was extremely humiliating. This key figure in the education and maintenance of order in the armed forces in essence did not exist. This was one of the reasons for the excessive number of officers, the hazing of conscript soldiers, and other troubles. Appeals and general demands will not do here. Regular sergeants are necessary who will serve under contract (first only master sergeants of companies or batteries, then deputy platoon commanders, and subsequently, whenever possible, commanders of sections, crews and teams). Other measures are also necessary to stimulate the moral and material interest of the sergeants in fulfillment of their duties. The professionalization of the army should begin with the sergeants.

The problem of a necessary increase in the personal responsibility of the soldier and the seaman is also urgent. In some armies which are based on general conscription, a soldier can legally be a most irresponsible person. Even a schoolboy can be kept down for another year, or expelled. The soldier and the seaman in such armies bear responsibility for their behavior only after committing a crime, and even then not always. If in society there existed the alienation of the worker from the means of production and from the land, in the armed forces there was the alienation of soldier and sergeant from the interests of military service. Along with improvements in education it seems necessary to take more effective measures of a disciplinary nature.

For soldiers of basic military service it might be possible to introduce pre-schedule dismissal (for good service), or the prolongation of service time by a few months (for careless behavior). For soldiers and sergeants of contract service, a system of penalties could be introduced, as exists in many armies. It should be clear, that if everyone is responsible for the soldier and he does not bear a stricter responsibility for his offenses, it will be difficult to expect a serious attitude from the soldier towards his service.

The issues of increasing the personal responsibility of military personnel are especially important for maintaining military discipline. The military first of all express through discipline their attitude towards the service. In this relation, a set of measures intended to change the attitude of personnel to service for the better, should play a decisive role.

Some armies still practice a completely uncivilized (from the point of view of logic and legality) way of estimating the state of discipline, by the number of incidents and convictions. (If a crime is unveiled, punished, and a criminal is taken to court, then discipline is bad; if the crime gets hidden, and gets passed over, then discipline is good.) Such an approach is the basic source of the tense situation in the armies and the navies, concerning the majority of investigative incidents. It constantly leads the officers into deceit and concealment of the violation of discipline which aggravates the situation. Discipline is a conscious and exact implementation of the requirements of the orders and regulations of the established military order. Only by this rule is it necessary to be guided on estimations of discipline.

Military service is needlessly complicated by far-fetched and obsolete requirements among the regulations which have existed since time immemorial. When the regulations are too bulky and contain many abstract rules, they are not usually obeyed. They cannot be obeyed, and consequently, create the preconditions for their constant violation and do not promote strengthening of discipline. In view of the above mentioned, it seems expedient to update military regulations in order that they support firm military order, and so that they do not burden military men, but are permeated with respect for the individuality of a military man, and provide protection of his legal rights. The barracks should not oppress soldiers and sergeants.

Regulations and other basic documents which regulate military service should be permeated with the idea of an organic combination of the principle of unity of command, necessary in military service, military order and organization – with a deepening and extension of the elements of democracy, certainly as regards the specificity of military service. The last condition is important too. The army is the army and it cannot be transformed into a trade union, otherwise it loses its sense of existence.

The American Bruce Parrot, of the Johns Hopkins University, an expert on the mutual relations of society and the army, considers that

> it is impossible to speak about a democratic army in the absolute sense. Therefore, it is unreasonable to take military decisions by vote, as well as to try to construct an effective military organization on the principles of democracy which exists in civil society.
>
> In this respect, a democratic army is nonsense. The democratization of the army is the creation of a system of guarantees of the rights and personal dignity of the person, which allows any military man subjected to humiliation or insult to easily appeal to civil authorities, to civil court; and people's deputies, in their turn, will have an opportunity to really protect and assert the rights of their military voters.[2]

It is difficult not to agree with such a statement.

In our opinion, it is desirable to carry out the perfection of the forms and methods of operational and combat training, in view of modern requirements, in the following directions.

Perfecting the Planning and Direction of Combat and Operational Training

First of all, it is necessary to determine the contents, and minimum necessary amount of educational measures. There are often disputes as to how many or how few exercises there should be. But the subjective, purely speculative approach, is not permissible here.

It is necessary to keep in mind that the improvement and mastering of all kinds of operations, and of various ways of fulfilling operational and combat tasks, is only possible along with the conducting of a certain amount of exercises and other measures on operations and combat training. The fact is that the training and conducting of modern operations covers a great span of diverse operational-tactical, military–technical, and special questions. First of all these questions apply to: putting headquarters and forces into a maximum degree of combat readiness; performing marches, railway, sea and air transport at great distances; conducting offensive and defensive combats and peculiarities of conducting combats in local wars and conflicts. Within the framework of each type of combat it is necessary to master by practice, ways of performing various tactical tasks: conducting meeting engagements, drawing or repulsing enemy counterattack, forced crossing of water barriers, questions of combat, logistic and technical maintenance, and many others.

These tasks cannot be mastered by practice during just one or two lessons or exercises, or even as a result of a yearly putting into practice of all measures of combat and operational training, let alone the necessity of the numerous recurrences of working out the most complex questions in view of the periodic replacement of personnel.

There is, therefore, the necessity for conducting a certain amount of educational measures, ensuring the necessary degree of battle coordination of units. For a reasonable approach to this question it is expedient to begin the planning of effective combat training by scoping educational tasks and listing the major operational-tactical questions that are necessary for mastery in practice by the organs of troop control; and then to determine a sequence of their study through various measures of effective combat training.

It seems expedient to plan for the mastery by practice of the basic questions of training and conduct of various kinds of combat over a certain period of time. Only under this condition it is possible to ensure a necessary combat readiness of units. Therefore it is desirable to carry out not only annual, but also prospective planning over two to three years.

In other words, it is desirable to coordinate whenever possible the basic measures of effective combat training (drills, military games, exercises) by a uniform operational-tactical and methodological plan, so that each training exercise involves the minimum of necessary issues, consistently fulfills the most important of them, and achieves a complete and all-round training of the organs of control and troops in the fulfillment of their tasks. These exercises should be carried out under combat conditions, as a result of conducting a system of interrelated measures.

While planning combat training, it is also expedient to make full use of the entire arsenal of forms and methods: training and methodological assemblies and monthly commander training; staff drills, exercises and war games; command and staff, special troop and tactical exercises. The command and staff and tactical exercises with troops are the most effective forms of combat training. But they achieve their goal only if they are preceded by the appropriate staff drills, staff exercises, war games and troop control drills, which means that the consecutive transition to the solving of more complex tasks is provided.

It has been suggested that the number of forms of training should be reduced by cutting out, for example, staff training exercises. But each exercise has its specificity, and when skillfully used supplements the others. For example, the staff exercises as against the command and staff or troop exercises, allow the commander to concentrate on the direct training of his headquarters personnel, without being distracted by the training of other subordinate headquarters or involved units. As the value of each exercise grows, it is necessary to carefully prepare headquarters and subordinated units for it, using all forms and methods, in order to pay primary attention in the exercise to the combat coordination of organs of troop control and units, without wasting time on their initial training, as often happens.

Lack of material resources necessitates a reduction in the number of exercises, especially when they involve large forces and a lot of equipment. But it is impossible to manage without a certain minimum. Therefore, even with the highest restrictions on the employment of motorized resources, it is desirable first of all to identify those operational-tactical tasks which are connected to the operational objectives of the formations, since mastering these tasks will require certain exercises, regardless of circumstances.

As is generally known, the number of exercises undertaken by the Russian army, the armies of the CIS, and eastern European countries, has recently been reduced by about one and-a-half to two times. Here and there their conduct has been stopped altogether. Exercises undertaken by NATO armies have decreased by 10–15 per cent. Nonetheless, NATO is not going to give up such large-scale exercises as 'Autumn Forge' (in 1992, 250 ships, 100,000 servicemen, and about 900 combat aircraft participated in this) and a number of corps and division exercises. Besides, the USA, pursuing its national interests, skillfully uses its actions in response to crisis situations in different regions of the world, for the training of its armed forces. The transportation, concentration, and combat use of massive groupings of its airforce, navy, and army in the region of the Persian Gulf in 1991, its large-scale landing operation in Somalia, its transport of forces to the Balkans and other regions, along with the accomplishment of political tasks, all provided combat training and coordination of joint actions of different branches of the US armed forces. These 'exercises' do not fall under the restrictions of the Stockholm agreement, and have been carried out partially at the expense of the UN, and various countries.

To reduce costs, the USA and other NATO countries widely practise computer war games and command and staff exercises at all levels. For example, in the exercise 'Certain Caravan' of the American forces in Europe, the automated battle conditions simulation system (ABCSS) was used, with the terminals deployed at the field and stationary control posts participating in the staff exercise. According to the press, 'the software of this system has allowed the performance of large-scale military actions with participation by about 200,000 personnel and more than 25,000 tanks and armored combat vehicles'.[3] In general, the wide use of computers and simulators during combat training provides great economy of motorized resources and ammunition.

Representatives from military-industrial complexes who have been involved in the introduction of computer equipment in the American army even declare that conducting exercises by computer simulation may render field command and staff and army exercises unnecessary. However, a closer and more critical study of the state of affairs has shown that the Americans, despite great achievements still have problems. The American computers used at the present are generally limited to the simulation of preprogrammed episodes and conditions, and react poorly to changes in the situation arising as a result of practical actions by forces. Combat terrain is poorly represented on displays and many other elements of conditions are neglected. When I visited military training institutions in the USA in March 1993, I saw that officers undergoing training on computer screens, have to update their working maps, since they have poor skills at finding their bearings on real ground.

When undergoing training on troop control by computer, it is enough to press the button and the second echelon of a brigade is engaged in action without any obstacles. In real conditions there will be many obstacles to overcome on the way to the attainment of this task. Therefore, such great experts in combat training as General W. Schoffner and F. Franks recognized that computerization of training, although being useful, must not become transformed into an end in itself. Computers can be a good addition to, but cannot completely replace, other forms of training. Therefore a reasonable combination of computer training and a minimum necessary amount of field exercises are required.

Methods of Developing Trainees Towards High Levels of Battle Skills and Military Art

The solution to this next problem rests on the correct understanding of the essentials within military science and military art, and the relationship between them. This question has a long history. Some commanders and military theoreticians have paid recognition to military science, others have not. Nevertheless, the majority of the most outstanding of them (Suvorov, Napoleon, Clausewitz, Lloyd, Moltke, the elder, Foche, Frunze and Zhukov) have seen in the phenomenon of war elements of both science and art, considering them as the two sides of military affairs. Marshals S.L. Sokolov and V.G. Kulikov have lately spoken in favor of such an approach.[4]

Military science comprises particular systematized knowledge about armed struggle; accumulated experience about military art is the ability to creatively apply this knowledge with due regard for particular conditions.

Historical experience shows, that with all the variety and complexity of wars, deep and essential internal relations are inherent in them, that they all have some common qualities which recur stubbornly and unavoidably. Therefore, the forms and methods of armed struggle are engendered not by the free will of great captains and commanders, but are dependent, like the actions of people, on certain law-governed patterns and specific principles. This explains the existence of military science, which represents a system of knowledge of the laws and military-strategic nature of war, the ways of its prevention, the construction and training of the armed forces, and ways of conducting armed struggle.

The other peculiarity of war, is that the law-governed patterns and objective phenomena inherent in it, being independent from the will and consciousness of people, do not act with the inevitable spontaneity of natural laws, but manifest themselves, like other social phenomena, through the actions of people. Knowledge of the laws, principles and ways of conducting armed struggle facilitates practical activity, and ability to forsee the progress of events, thus facilitating conscious actions. But this knowledge cannot provide the answer to the question of how to perform in one or another particular situation. Therefore, the rules of military science cannot be applied in all cases, and irrespective of all conditions with the same constancy and identical outcome as opposed to the laws of the natural sciences.

Furthermore, the question is also complicated by the necessity of interacting with phenomena and law-governed patterns, which as often as not interlace and resist each other, creating the most inconsistent situations in the course of armed struggle. Each side of a conflict aspires to use objectively formed circumstances to its own advantages, and to impose its will on the opponent, to hide its intentions, to deceive the resisting side, to stun it with unexpected actions.

All this creates extremely complex, intense and continuously varying conditions during armed struggle. Therefore, for successful military actions it is not sufficient to know only the foundations and principles of armed struggle and military science. Military art is required too. If victory and defeat in war could be determined by the automatic actions of objective phenomena and law-governed patterns and if the fundamentals of military science could be applied in practice, like the laws of mathematics or physics, military art would no longer make sense, and such a concept would cease to exist. In reality, military art is essential; the outcome of armed struggle very often depends on how precisely military leaders and commanders at all levels are able to assess the objectively formed circumstances, to make expedient decisions, and to implement them skillfully in practice.

One of the participants in discussions on this issue in Russia, Y. Smyslovsky, wrote: 'the study of any science does not at all guarantee mastering the art directed by this science, life displays numerous examples of so-called "learned fools" who, possessing broad scientific knowledge to perfection, appear completely incompetent in the field of the corresponding art. It is one thing to know, but an entirely different thing to be able'.[5] The whole of military history testifies that military science cannot offer ready-made solutions, it can only serve as a reference point in dealing with reality.

To keep military art on a high level, it is necessary not only to be proficient in science, but also to possess an advanced level of operational-tactical thinking along with a creative approach; to be able to expeditiously evaluate a situation and to analyze it, to possess high organizational and moral-combat qualities such as courage, boldness, resolution, initiative, independence, firmness and persistency in reaching one's goals. These qualities are not inherent or peculiar exclusively to gifted people. They also do not appear as a result of the study of theory alone, but rather during combat and effective training, in practical activities, and in the bulk of military service in peace and wartime. There is a considerable distance between the mastery of theoretical knowledge, and the ability to apply it in practice. For example, in the Soviet army at the beginning of World War II there was probably no military leader or commander who did not understand theoretically, and was not aware from the experience of the past, of the necessity of the concentration of main efforts on the decisive axis, the creation of strike groupings, scrupulous reconnaissance, and reliable destruction of the opponent with firepower. Nevertheless, it took a lot of time, considerable effort and many losses before it became possible to take possession of the art of solving these and other tasks.

Military art starts at the point where, on the one hand, deep theoretical knowledge and its creative use help the commander to see the mutual relations between the occurring phenomena, and to have confidence in the situation; and on the other hand, in situations where the commander is not held down by general theoretical schemes, and aspires to penetrate deeper into the essence of the situation, enables him to catch its favorable and unfavorable peculiarities, and proceeding from their analysis, to find original decisions and methods of actions, which in the greatest degree correspond to the combat task. The maximum degree of conformity of decisions and actions of commanders and forces to specific conditions, is the major law of military art, as it exactly expresses its essence, determining the most essential and steady relations, parity of objective and subjective factors, internal motive forces and the basic causes of victories and defeats in war.

During the postwar years, particularly among the armies of the Warsaw Treaty Organization (and contrary to the experience of war gained at such a high cost, formally fairly, but in essence unilaterally), the importance of theoretical knowledge was emphasized, and the necessity of the cultivation among officers of a number of qualities without which military art is impossible, was underestimated. This was formally justified, but, in essence, one-sided. During training and exercises, there were often no situations in which trainees would be compelled to reveal these qualities. Therefore, the correct understanding of the nature and cooperation of military science and the practice of military art as two interconnected and interdependent parts of military affairs has not only methodological, but also great practical importance, especially from the point of view of the direction, forms and methods of military training, and the education of military personnel.

Even now in many armies the basic flaw in the preparation of many officers is the tendency toward schematic solutions and almost complete detachment from the concrete situation – the formalistic, simplified approach to the solution of many issues of troop control, when much is done not because of the demands of the given combat conditions but because it is 'supposed to be done that way'.

Therefore, in the education of officers, along with theoretical knowledge, special attention must be paid to the development of organizational skills, strong will, and, in general, practical skills of troop control. As experience shows, the most advanced theoretical views become a material force and a component of military art, if they are mastered practically by the greater part of the officer corps. During World War II line officers of all ranks always learned the art of conducting a battle and an operation. Such study before performing every new combat task, was carried out in the Soviet troops in Afghanistan, and in the coalition forces before the beginning of war in the Persian Gulf in 1991.

Why did many offensive operations of the Soviet Armed Forces in 1941–42 not meet with success? One of the causes was that senior commanders, commanders and their staff did not take into account in making their decisions changes in combat conditions, limited their functions to the taking of the decision and the assignment of tasks – and then not always in a timely fashion – and paid little attention to organizational work with troops in preparation for combat. Specificity, efficiency, and thoroughness in the organization and support of combat actions was lacking in the work of commanders and their staff. And if this does not exist, then the best laid plans and the most brilliant exposition of their content lose any sense.

Taking the above into account, it is necessary to work within the exercises all issues relating to the practical organization of the interaction as well as of general organization of combat, and its combat, technical and logistic support in all their details. In exercises as much time should be allocated for the accomplishment of these tasks as is required in combat conditions. Even in command and staff exercises, where only nominal units are involved, it is necessary to conduct all planning measures in preparing for combat: to actually organize and conduct reconnaissance, to accomplish engineer preparation of the area, to create prescribed material reserves, to carry out regular maintenance service of combat equipment and so on.

Special attention should be paid to organizing the interaction of sub units and units of all combat arms. This work should not be reduced to the decision of formal theoretical questions, as sometimes happens in exercises, but should bear a practical character, and should, first of all, deal with fundamental issues, upon the correct solution of which the cooperation of the forces depends; it should provide uniform understanding of combat tasks and methods of their fulfillment by all personnel. The commander of the exercise must carefully check that everyone accurately understands the directions of the forthcoming actions, where the fire means of the 'opponent' are located (coordinates of the targets), which artillery or aviation and when, is to hit them, the order and signals for calling down and ceasing fire during combat, the location of passages through the obstacles of the 'opponents', and other questions. When necessary, one should demand additional or repeated mastering of these questions, but not allow under any circumstances that commanders and headquarters under training begin an insufficiently prepared combat or operation.

Hence, the formation of qualities necessary for the display of a high level of military art, is first of all reached by strictly putting into practice the main requirements of the training and military education of forces – to teach those subjects which are required under combat conditions.

Further Developments in the Training of Headquarters and Forces, to Ensure Maximum Approximation of Training to Battle Conditions

Independent study on his own by each general, admiral and officer, remains the most important major form. For the military man war is a test. It is not always known when this is going to be held, but it is necessary to prepare for it throughout his entire life. However, independent study cannot be the responsibility of the officers themselves. For this purpose, on the one hand, the inculcation of a strictly critical attitude among officers toward their training – especially for line officers – and the creation of highly exacting conditions and intolerance towards any smudginess and complacency, are required. The independent study of an officer should be skillfully directed by senior commanders through individual tasks, interviews and other methods. The experience of some armies deserves attention, where annually the entire officer corps is given a list of books to read. On the other hand, further increase of responsibility and interest of the officers in the results of effective combat training, and its importance in the system of officer evaluation and promotion in service is necessary.

Experience shows that it is expedient to maintain consistently the principle of conducting command and staff and other exercises, as a rule, under the direction of senior commanders for effective combat training of headquarters and forces. Combat conditions always contain some elements of uncertainty. If a commander knows everything beforehand, the exercise loses its sense. When conducting exercises independently, the commander and other authorities who know the situations for both sides fail to get practice in troops control, which is necessary for specific combat conditions. For example, there is no need to conduct reconnaissance, and other actions acquire a purely formal character. This was the main flaw of the joint Russian–American exercise in 1994 at Totskoye.

Exercises with troops remain the highest form of operational training. In the past and sometimes nowadays, disputes take place on which is more important: well-trained companies and battalions, or a high level of training of the organs of troop control at the operational-strategic level. Experience of World War II has convincingly shown that both are important. During training exercises of the Red Army in the 1930s, the commanders and headquarters of military districts, and sometimes of armies, acted as directors of the exercises and, therefore, failed to get necessary practice in troop control in the conduct of operations. After the Soviet–Finnish war, the main emphasis was placed upon combat training of units and operational exercises with troops. Maneuver exercises were not conducted at all; the training of headquarters was limited to command and staff exercises and war games. But this turned out to be insufficient. When the war began, it was not the companies or battalions that were least prepared, but the organs of troop control at the operational-strategic level. Life has shown that for high-grade training of commanders and operational headquarters, and for the more complete practical improvement of the interaction of forces and the combat coordination of formations, it is necessary to carry out operational exercises with troops, where commanders and staffs control real troops. Senior headquarters have no less complicated duties in troop control than platoon or battalion commanders, and they should constantly prepare for their fulfillment too.

The creation of an instructive initial situation, its escalation in the course of the exercise and simulated combat performance, all have especially great importance for the effective conduct of an exercise.

First of all, the initial conditions need to correspond to the nature of combat, to operational-tactical plans, and should put those being trained within the complex conditions of ground, air, sea and radio-electronic situations. The initial situation should create objective conditions for the actions of sides, should abound in difficult situations requiring serious reflection and creative solutions, and subsequent original course of action. Trainees should use their skills to find a way out of the most difficult situations. Combat training conditions should exclude obvious, unequivocal decisions, and should provide an opportunity for taking various courses of action and for choosing the most appropriate among them.

As often as not, it is possible to hear during discussion of exercises some fair complaints that some commanders and staffs pay little attention to questions of reconnaissance, or to technical and logistic support. But this frequently occurs because in the course of the exercises no conditions are created that require decisions and actions on these matters (data on the opponent comes from directors and umpires of the exercise irrespective of the practical organization and conducting of reconnaissance, electronic jamming is poor, troops have no shortage of ammunition or other means, and so on). There is criticism of poor maneuvering, linearity of actions of troops, and insufficient initiative, but directors and umpires of training exercises do not always create conditions that promote the practical execution of all this. But if we want to develop certain qualities in commanders, staffs and troops, it is necessary to put trainees within conditions where they could display these qualities.

Under the improved dynamics of combat, and the demands set on modern battle with the participation of combined arms and services according to their peculiarities, as considered above, the following points should be used as a basis: effective reconnaissance, validity of the decisions taken and timeliness of assigning tasks to subordinates; reliable fire destruction of the opponent, initial destruction of enemy precision weapons-systems and skillful use of these results; full utilization of forces for destruction of the opponent and fulfillment of the tasks assigned; surprise and resolute actions; firm troop control and continuous interaction of units of all combat arms; preserving high morale among personnel, skillful and timely combat, logistic and technical support.

In view of all the above, it is also very important correctly to assess drills and exercises in military training institutions and with troops. The fact is that a commander who subjects trainees to more complex conditions during a drill or an exercise puts the trainees in a less favorable situation when they are prone to making more mistakes than do the officers and troops which train under structured conditions. Some senior commanders only pay attention during exercises to the actions of commanders and troops, ignoring the conditions in which they are placed. This accounts for the tendency to simplify conditions, and even prompt the trainees to guess the decision of the command, and to be afraid to make a mistake. Thus, it would be more correct to estimate the quality of a drill or exercise not only by the actions of those being trained, but also by their methods of instruction, and the degree of approximation of training situations to combat conditions. There will always be more mistakes under complex conditions, but they can always be eliminated. The main thing is that real combat training be held.

Thus, operational and combat training management, especially the preparation and conduct of exercises, is becoming more and more complex, requiring both a good command of military art and knowledge of methodology.

On the whole, an appreciable increase in quality of operational and combat training will in the future play a more important role within the system of the intensification of combat effectiveness, and the combat readiness of the army and navy. Higher levels of effectiveness of their actions will be reached mainly by the reorganization of the contents of military training in accordance with the changing character of armed struggle. The forms and methods of education should provide the maximum approximation to combat conditions. For this purpose, it is necessary to place the latest achievements of pedagogy and psychology into the basic process of education, and to make use of computers, simulators, and other means of simulation.

All measures of operational and combat training should be carefully prepared and skillfully conducted, so that they may really meet the purpose and directly affect the increase of combat effectiveness of control organs and troops.

NOTES

1. *Militarizm i sovremennoe obshchestvo (Matrerialy nauchnoy konferenstii v Institute SshA i Kanady Akademii Nauk SSSR)* (Moscow, 1987) p. 130.
2. B. Parrott, in *Voyennaya mysl'*, No. 12 (1992) p. 61.
3. *Krasnaya zvezda*, 20 April 1993.
4. S.L. Sokolov and V.G. Kulikov, in *Voyennaya mysl'*, special issue (1985) pp. 4–15; No. 12 (1993) pp. 81–8.
5. Y. Smyslovsky, in *Voyennaya mysl' i revolyutsiya*, Vol. 3 (1922) p. 19.

CONCLUSION

In our complex contemporary world, positive changes are taking place connected with the end of the Cold War and former global confrontation. These changes facilitate the peaceful settlement of international problems. It is in the interests of all humankind to attempt to eradicate war, and there is nothing more noble or essential than the struggle for peace. But in reality, a large number of acute political, economic and ethnic conflicts still persist, and it is not always possible to settle these by peaceful measures. This is why so far it has been impossible to eliminate war, and a peaceful era on earth will probably not come soon.

The destruction of the old bipolar world disrupted the world's strategic stability however fragile, unreliable, and sustained with difficulty; and now in the new world order, the mechanisms for ensuring the stability of this multipolar world are not yet devised. In general, the world has become even less secure. Under these conditions, the key task required of the international community is to coordinate efforts towards the political prevention of wars, and when this fails, to implement immediate action towards the interdiction of any aggression.

The gloomy years of the Cold War and the logic of development of events have brought the world community, and the majority of politicians and military leaders to an appreciation of the senselessness of nuclear strategy, the impossibility of the massive application of nuclear weapons, or the conduct of nuclear war. Separate instances of the unauthorized employment of nuclear weapons, their use by terrorists, or use for provocative goals could occur. But in general nuclear weapons, while they are retained, remain a means of strategic deterrence, influencing the attainment of political and strategic purposes.

New dispositions of military–political forces in the world make even global war using conventional weapons less probable. But it cannot be excluded altogether, because history has shown a number of examples when war broke out for some objective reason, although neither side wanted the war. At present, and in the long term, local wars and conflicts capable of growing into large-scale military confrontations are becoming more widespread and dangerous.

The suggested prospects for the development of the military–political situation will largely shape the direction of military construction, when the great powers will have to ensure readiness of their armed forces for participation in local wars, military conflicts, and international peace-making actions. At the same time, the foundations of the structure of armed forces in peacetime, their composition, mobilization, and deployment capability in war, will all be laid down with a view to ensuring their readiness to rebuff aggression and to participate in large-scale warfare. The primary direction for the combat training of armies and naval forces will be preparation for conventional warfare, under the constant threat of nuclear escalation. The use of armed forces in future war will be of a multi-variant nature, requiring greater flexibility and diversity of actions, and a combination of direct and indirect operations with an increasing role for the latter.

Tendencies within the development of warfighting methods are contradictory in nature and have no unique direction. On the one hand, the most radical political and strategic goals can be set, but on the other hand, the new weapon systems delimit themselves and render these goals unattainable, predetermining the framework of the armed struggle. Modern high-precision weapons enable the prompt destruction of large hostile groupings, but counteraction by the same means, along with enhanced replenishment capabilities, might result in comparatively protracted efforts and stressing of all available forces and means for accomplishment of the assigned missions.

The tendency for the domination of offensive armaments, and in general offensive capabilities, is still preserved. At the same time, tanks are counteracted by more sophisticated anti-tank means; aviation by air defense facilities; missiles by electronic warfare; and increased fire power of the offensive by the ability of defense to withstand it. Not only offensive, but also defensive operations have become air–land in nature.

The enhanced fire power and the susceptibility of troops to it demand dispersed deployment and actions of troops and naval forces. Simultaneously, objective laws and their resulting principles can be applied, as, for instance, the principle of the uneven distribution of forces and means, which necessitates certain actions for ensuring success, including finding new ways for the concentration of major efforts on the accomplishment of main tasks in combat and operations.

New reconnaissance capabilities, especially space-based systems, seemingly make the dispositions of forces of both sides 'transparent', but as it was demonstrated in the Falklands and in the Gulf area, new camouflage facilities in combination with efficient tactical masking and misinformation (variants of each method in a specific situation are virtually unlimited) open up new opportunities to provide secrecy and surprise.

The sophistication of warfighting methods, and the necessity to coordinate their application, require computers and other means of automated control to make centralized command and control possible and expedient. The variability of situations, with the fluidity and dynamism of the developments of events, make effective command and control possible only provided that relative independence and initiative is entrusted to each level, which in its turn requires a fundamental transformation not only of means, but also of methods and psychology of control activity, necessitating new requirements in the training of military personnel.

The disappearance of the global bloc confrontation among states, the primary threat of local wars and conflicts, advanced economic possibilities of leading countries, and the elaboration of weapons and material predisposes priorities towards a transition to professional armies. Demands for greater integration of society and the army under new socio-political conditions, an increase in the educational level of the young, new control devices, and hence possibilities for training within shorter terms, as well as some other considerations, all make it illogical to man all positions with professionals, and so necessitate the retention of elements of general conscription.

That is why for an adequate profiling of future armed struggle it is not sufficient to view only separate and observable phenomena, and one-sidedly to use them as guidelines. One should take into account contradicting tendencies in the development of military affairs, and draw conclusions from the array of these complicated, many-faceted phenomena, for the military use of future power. This, while being one of the major lessons of history, is at the same time a vivid reflection of the tendencies and prospects in the evolution of military affairs.

Wars of the future might present an even bigger challenge for peoples and armies than previously. Realistic estimations of the prospects of development in military art, and the nature of future armed struggle devoid of ultra-radical or conservative obsessions, is a prerequisite for the effective (with minimal losses) solution of defensive tasks by armed forces of peace loving nations. Only on these conditions can commands and headquarters at all levels within ground forces, navy and aviation actually be able to prepare for whatever might be required of them on future battlefields.

INDEX